Darwinism pollutes the Word of God

Darwinism teaches that death existed in the animal kingdom for millions of years before the first man ever appeared on the scene. In fact, Darwinism teaches that millions and millions of years of death and suffering are the only means—the magic ingredient—whereby mankind could have somehow mutated his way into existence.

But this flies in the face of Biblical Creation, which teaches that God created the world perfect, and that it was man's original sin which caused death to enter God's perfect creation, thus separating us from our Creator.

Two opposing viewpoints.

Two VERY different implications—for if original sin never separated us from God, bringing in the curse of death upon the world, then why would there have even been a need for Jesus Christ to go to the cross to receive the death penalty for our sins?

By undermining people's faith in Biblical Creation, which includes the accounts of Adam and Eve's original sin, the resulting separation from our loving Creator and the first promise of our coming Redeemer, Darwinism destroys the very foundation of the Gospel of Jesus Christ!

The damage already done to societies around the world by Darwinian philosophy is almost too deep to comprehend and it is almost too late to turn the tide. However, there is a vast difference between *too late* and *almost too late*. So our challenge to you is to read this report, and then pick up your cross and make a difference in this world.

The war is not against flesh and blood, but against powerful forces of darkness.

We hope that this book will lead you to seek a way to make a difference.

Please visit www.Creationministries.org and prayerfully consider involvement with our **Make This Your Own DVD Ministry**. This is one of the easiest, least expensive and most powerful ministries God has provided for those willing to join us as foot soldiers in this war.

Enlist now, and help us turn the tide.

God bless.

Russ Miller

UCS PRESS PMB 119
1702 W. Camelback Rd. # 13
Phoenix, AZ 85015

UCS PRESS is an imprint of MarJim Books, Inc.

Bible verses and passages quoted in this report are
from the King James Version.

Re-titled edition, second printing, June 2010

NOTE: This revised report formerly was titled
The Facts Are Talking, But Who's Listening?
with first printing in December 2008.

Printed in the United States of America.

ISBN: 978-0-943247-96-0

Dedication

To the growing army of
"Make This Your Own DVD Ministry"
foot soldiers valiantly exposing the lies of Darwinism.

Acknowledgments

Special thanks to our wives
Joanna Miller and Marty Dobkins for their
unconditional love and support, and to
Valeri Marsh and Jim Porter
for their continued insightful editing assistance.

Darwinian Delusion

Russ Miller
with Jim Dobkins

About the authors

After spending 18 years building a successful nationwide management recruitment firm, Russ Miller walked away from it all in the year 2000 so he could found Creation, Evolution & Science Ministries. He's dedicated his life to studying Creation-Evolution issues and developing CESM's popular, yet challenging, PowerPoint programs. Russ has presented over 1,000 seminars and church service messages and appeared on international telecasts. Through thousands of radio programs he has presented scientific evidences which have challenged hundreds of thousands of Christians worldwide to confidently believe the whole Bible – word for word and cover to cover. He lives with his wife Joanna in Flagstaff, Arizona.

Jim Dobkins is a Southwest-based writer. His co-author credits include *Winnie Ruth Judd: The Trunk Murders*, *Machine-Gun Man*, *The Ararat Conspiracy* and *My Boss was the BTK killer*. He conceptualized and wrote the dramatized short documentary *Someone Who Cares* and has ghostwritten several books. He has written over 1,000 articles. Primarily known as a crime writer, he now focuses on writing inspirational books and collaborating with Russ Miller in developing books in *The GENESIS Heritage Report Series*. This is the second report in this series, following *The Theft of America's Heritage*. Dobkins resides with his wife Marty in Phoenix, Arizona.

Contents

Contents continued

Contents continued

Key Words: Definitions and Abbreviations

Darwinism: As a theory, Darwinism holds that, through long ages of naturally-occurring death, suffering and struggle, life forms have evolved from a common ancestor due to natural selection removing the lesser of the species, leaving improved variations to take over the population.

Neo-Darwinism: A theory which holds that, through long ages of naturally-occurring death, suffering and struggle, life forms have evolved from a common ancestor – here is where the difference between Darwinism and Neo-Darwinism sets in – AFTER random genetic mutations added new and beneficial genetic information to an existing gene pool and natural selection removed the lesser of the species, causing the mutant to take over the population.

NOTE: In this book, unless otherwise specified, both of the above terms are included in our usage of the term Darwinism.

Secular Humanists: Those who espouse the atheistic religious worldview that "millions of years of naturally occurring evolution" brought all life forms into existence. They preach Darwinism as their gospel, and that God is imaginary. They orchestrated the removal of prayer and the teaching of Biblical Creation from public schools, replacing these Biblical foundations with the teaching of their religious worldview. They control public education at all levels in America, pushing naturalistic Darwinism with relentless religious fervor.

Intelligent Design (ID): A theory founded on the concept that many living, as well as inanimate objects in the universe, are so complex that they must have been created by some form of superior intelligence, and not necessarily by the God of the Bible.

Intelligent Biblical Design (IBD): A theory totally supportive of the creation account recorded in the Book of Genesis, that living, as well as inanimate objects in the universe, are so complex that they had to have been created by the Biblical God revealed in the Holy Bible.

Devolution: The concept that all life forms including man have devolved rather than evolved following Adam and Eve's sin of disobedience which resulted in their banishment from the Garden of Eden as reported in the Book of Genesis.

Science: Knowledge honestly derived from the thorough and objective examination of available and observable evidence.

Prologue:

Amazing example of Intelligent Biblical Design

As you'll read in this report, the real focus should be on our Intelligent Biblical Designer, and not on merely the concept of Intelligent Design.

Here's one example of the fingerprints of our Intelligent Biblical Designer found throughout His Creation.

The bucket orchid

The petal of a Bucket Orchid has a slimy surface which causes a bee, attracted by the aroma of nectar put out by the plant, to slip and fall into the orchid's "bucket." In the bottom of the bucket is a pool of liquid and the only way for the poor little bee to escape from the bucket is through a tunnel that goes from the side of the pool to the outside of the flower. There is even a step at the edge of the pool for the bee to climb up onto and go into the tunnel.

However, as the bee crawls through the tunnel and towards freedom, the walls of the orchid contract and capture it, holding the insect while the flower glues two pollen sacs to the creature's back.

After allowing time for the glue to dry, the flower then releases the bee so it can fly off on its search for nectar and if it lands on another Bucket Orchid the entire process is repeated. The bee slips and falls into the pool then crawls up the step and into the tunnel where the flower walls contract on it again. I am fairly sure that the little fellow is thinking, *Déjà vu! Wasn't I here before?*

However, this time two hooks come from the orchid and remove the two pollen sacs, completing the pollination process before releasing the insect to go about its business.

WOW!

Talk about awesome proof of our Intelligent Biblical Designer!

Philosopher Malcolm Muggeridge stated in the Pascal Lectures, University of Waterloo, Ontario, Canada, that:

"...the theory of evolution...will be one of the great jokes in the history books of the future. Posterity will marvel that so flimsy and dubious an hypothesis could be accepted..."

ONE

The Biggest Con Game On The Planet

You will find no pussyfooting around about the Creation vs. Evolution debate in this report. I've been on both sides of the fence, having previously been a Theistic Evolutionist. I didn't have any problems believing that the Biblical God created the universe, but I was hung up on the notion that He had to have taken *millions and billions of years* to get everything done.

Evolutionism was wrapped around my brain like a boa constrictor, gradually squeezing the truth out of me. The seeds were planted during my grade school years by science and biology textbooks that presented Darwinism as science, while dismissing the Bible as a fairy tale.

Darwinists would've loved me. I didn't do or say anything to keep others from falling for the lies that are force-fed to students at all levels of public education in America and in many other countries. Such a deal they have with the textbooks and progressive education controlled by Secular Humanists.

Students only hear one side of the story.

And if teachers or professors even utter out loud the possibility of an opposing opinion they risk getting fired.

Whatever happened to the supposedly guaranteed First Amendment right of free speech?

Secular Humanists have trampled on the First Amendment; in fact, on the whole United States Constitution. And on the concept of a Biblical Creator that is cited numerous times in the Declaration of Independence. The very Creator who endowed Americans with certain unalienable rights.

Several generations of students have not been taught the truth that the Republic of the United States of America was founded upon predominantly Christian principles by predominantly

Christian men. We thoroughly documented this in the first report in The GENESIS Heritage Report Series: **The Theft of America's Heritage**.

Until 1998 I didn't give a hoot about the Creation vs. Darwinism debate. Then I watched a video made by a Creationist. For the next two years I searched out and devoured everything I could about science, and what Darwinists were preaching as opposed to what the Bible proclaims.

If you read the blurb about me in the **About the authors** segment on the page right before **Contents** you know that I gave away a lucrative business in 2000 and founded Creation, Evolution & Science Ministries. My wife Joanna and I joined the army of foot soldiers for truth full bore and have never looked back. Our mission is to subject the claims of Darwinism to the scrutiny of the observable facts – real science – which is an enemy to Darwinism, in an honest way that will allow people to decide for themselves what they want to believe.

You will read the truth in this report.

The claim that Darwinism is science is a lie, a fraudulent bluff, a hoax, a sham, an all-out scam used to get billions of people to buy into the supreme lie that the Bible is a fairy tale and that it took *millions and billions of years* for our universe to reach the highly-evolved state that we see today.

We will reveal the truth about this and other Darwinian claims in this report as we present dozens of examples that refute Darwinism.

And, yes, absolutely, I will even show you how to refute Darwinism in seven seconds flat – and that's when I'm not talking fast. If you can't wait to find out how this can be done, you have my permission to sneak a peek at Chapter Seven: **Seven Seconds Flat!**

C.S. Lewis wrote so beautifully in his classic essay *Weight of Glory* that:

"To be ignorant and simple now . . . would be to throw down our weapons and to betray our uneducated brethren who have, under God, no defense but us against the intellectual attacks of the heathen."

TWO

It Was Very Good

The ultimate manufacturer's stamp of approval came at the end of six days' labor during which God created the universe and everything in it. He looked at what He'd made and, as Genesis chapter one, verse 31 declares, "…behold, it was very good."

This first chapter in the Bible is elegantly simple, laying out step-by-step the order in which God manifested His creation. Yet this chapter is among the most loved and despised of all Scripture passages.

To Christians who believe the Bible word for word and cover to cover it is the answer to the question, "When and how did life begin?"

To Secular Humanists, who worship naturalistic Darwinism, it is nothing more than a fairy tale fabricated to convince people that there is a Creator who has set down both moral and spiritual guidelines to which they will be held eternally accountable.

Two worldviews are at war with each other:

A. The Biblical worldview, which explains that all things come from our Creator God.

B. The Darwinian or Secular worldview, which tries to explain all things by naturalistic processes, without God being involved.

It is important for background and memory refreshing that you read the 31 verses of Genesis, chapter one. And if you have never read this chapter, this is your opportunity. Here are those 31 verses that generate such great passion and anger, depending upon which worldview is embraced:

[1] In the beginning God created the heaven and the earth.

[2] And the earth was without form, and void; and darkness was upon the face of the deep. And the Spirit of God moved upon the face of the waters.

[3] And God said, Let there be light: and there was light.

[4] And God saw the light, that it was good: and God divided the light from the darkness.

[5] And God called the light Day, and the darkness he called Night. And the evening and the morning were the first day.

[6] And God said, Let there be a firmament in the midst of the waters, and let it divide the waters from the waters.

[7] And God made the firmament, and divided the waters which were under the firmament from the waters which were above the firmament: and it was so.

[8] And God called the firmament Heaven. And the evening and the morning were the second day.

[9] And God said, Let the waters under the heaven be gathered together unto one place, and let the dry land appear: and it was so.

[10] And God called the dry land Earth; and the gathering together of the waters called he Seas: and God saw that it was good.

[11] And God said, Let the earth bring forth grass, the herb yielding seed, and the fruit tree yielding fruit after his kind, whose seed is in itself, upon the earth: and it was so.

[12] And the earth brought forth grass, and herb yielding seed after his kind, and the tree yielding fruit, whose seed was in itself, after his kind: and God saw that it was good.

[13] And the evening and the morning were the third day.

[14] And God said, Let there be lights in the firmament of the heaven to divide the day from the night; and let them be for signs, and for seasons, and for days, and years:

[15] And let them be for lights in the firmament of the heaven to give light upon the earth: and it was so.

[16] And God made two great lights; the greater light to rule the day, and the lesser light to rule the night: he made the stars also.

[17] And God set them in the firmament of the heaven to give light upon the earth.

[**18**] And to rule over the day and over the night, and to divide the light from the darkness: and God saw that it was good.

[**19**] And the evening and the morning were the fourth day.

[**20**] And God said, Let the waters bring forth abundantly the moving creature that hath life, and fowl that may fly above the earth in the open firmament of heaven.

[**21**] And God created great whales, and every living creature that moveth, which the waters brought forth abundantly, after their kind, and every winged fowl after his kind: and God saw that it was good.

[**22**] And God blessed them, saying, Be fruitful, and multiply, and fill the waters in the seas, and let fowl multiply in the earth.

[**23**] And the evening and the morning were the fifth day.

[**24**] And God said, Let the earth bring forth the living creature after his kind, cattle, and creeping thing, and beast of the earth after his kind: and it was so.

[**25**] And God made the beast of the earth after his kind, and cattle after their kind, and every thing that creepeth upon the earth after his kind: and God saw that it was good.

[**26**] And God said, Let us make man in our image, after our likeness: and let them have dominion over the fish of the sea, and over the fowl of the air, and over the cattle, and over all the earth, and over every creeping thing that creepeth upon the earth.

[**27**] So God created man in his own image, in the image of God created he him; male and female created he them.

[**28**] And God blessed them, and God said unto them, Be fruitful, and multiply, and replenish the earth, and subdue it: and have dominion over the fish of the sea, and over the fowl of the air, and over every living thing that moveth upon the earth.

[**29**] And God said, Behold, I have given you every herb bearing seed, which is upon the face of all the earth, and every tree, in the which is the fruit of a tree yielding seed; to you it shall be for meat.

[**30**] And to every beast of the earth, and to every fowl of the air, and to every thing that creepeth upon the earth, wherein there is life, I have given every green herb for meat: and it was so.

[**31**] And God saw every thing that he had made, and, behold, it

was very good. And the evening and the morning were the sixth
day.

At various places in this report one or more of these verses will
be referenced during particular points of discussion.   Scripture
from other books in the Bible will also be cited when appropriate.

Now dig in and prepare to experience the Creation vs.
Darwinism debate like you probably never have before.

Here is an observation about the suppression of anti-evolution sentiment made over 50 years ago:

"As we know, there is a great divergence of opinion among biologists, not only about the causes of evolution, but even about the actual process. This divergence exists because the evidence is unsatisfactory and does not permit any certain conclusion. It is therefore right and proper to draw the attention of the non-scientific public to the disagreements about evolution. But some recent remarks of evolutionists show that they think this unreasonable. This situation where scientific men rally to the defense of a doctrine they are unable to define scientifically, much less demonstrate with scientific rigour, attempting to maintain its credit with the public by the suppression of criticism and the elimination of difficulties, is abnormal and undesirable in science."

Dr. W. R. Thompson in his introduction to *The Origin of Species* by Charles Darwin; J. M. Dent and Sons, 1956, page xxxii.

THREE

It's Really Devolution

My co-author maintains that the term evolution is a gross misnomer, and I wholeheartedly agree. He says that the correct term for what has been happening since Adam and Eve were expelled from the Garden of Eden is devolution.

It is the perfect word to describe what has happened since Adam and Eve committed the original sin when they ate the fruit of the tree God had warned them not to eat in the garden.

Until Adam and Eve's act of disobedience, they had lived in a perfect estate. There was no death, and no suffering. Everything they needed, including the companionship of their Creator, was there. There were not even thorns or weeds to mar the view in the most beautiful environment in God's Creation.

They were created to be immortal. Yet they were totally human, which meant they were subject to temptation as all of us humans are today; and even as Jesus Christ was during His 40 days of fasting in the wilderness during which He was tempted by Satan.

While Jesus Christ resisted the temptations of Satan, Eve did not resist the charms of the serpent, whose cunning words recorded in the third chapter of the Book of Genesis inflamed Eve's lust to eat of the fruit of the tree of knowledge of good and evil:

> [1] Now the serpent was more subtil than any beast of the field which the LORD God had made. And he said unto the woman, Yea, hath God said, Ye shall not eat of every tree of the garden?
> [2] And the woman said unto the serpent, We may eat of the fruit of the trees of the garden:
> [3] But of the fruit of the tree which is in the midst of the garden, God hath said, Ye shall not eat of it, neither shall

ye touch it, lest ye die.

[4] And the serpent said unto the woman, Ye shall not surely die:

[5] For God doth know that in the day ye eat thereof, then your eyes shall be opened, and ye shall be as gods, knowing good and evil.

[6] And when the woman saw that the tree was good for food, and that it was pleasant to the eyes, and a tree to be desired to make one wise, she took of the fruit thereof, and did eat, and gave also unto her husband with her; and he did eat.

Then what often is referred to as *Adam's Fall* triggered a devastating series of events:

Adam and Eve were expelled from the garden.

The ground was cursed so that Adam would have to toil hard all of his life to provide food for his family.

Eve and all women descendents of hers bore the curse of pain during childbirth.

Most catastrophic was that death and suffering entered into the world that had been perfect – until Adam and Eve committed the original sin.

This last point is the great divide between those who believe the Bible word for word and cover to cover, and Darwinists or their allies. Darwinian beliefs are based on *millions of years* of death and suffering before modern man evolved from primordial goo and through various lower life forms, until finally emerging from some apelike ancestor into the human form we exist in today.

There is no common ground between Darwinian-style evolution (one kind of creature evolving from another kind of organism) and Biblical Creation.

They cannot both be true. Either one is true and the other is false, or they are both not true.

As I mentioned in the first chapter, I make this statement as one who used to be a Theistic Evolutionist – I believed that God used evolution as a tool while He created the universe over *millions and billions of years.* I was comfortable with this king of excuses, this horrendous compromise of God's Word, until I heard a Creationist speaker. Then over the next two years I devoured everything I could about Darwinism and what the Bible declares about how the universe and everything in it came into existence.

Here's the bottom line:

If the *millions and millions of years leading to Darwinian evolution* Secular Humanists base their naturalistic notions on is true, the Holy Bible is not true.

But if the words God inspired men to write down in the Holy Scriptures are true, then the *millions and millions of years* is one huge lie.

God either created the universe like He said He did in His love letter to mankind – the Holy Bible...

Or He didn't.

God either judged the world with a flood that killed every human on the earth, except the eight people protected on the Ark, like His Word proclaims in the book of Genesis...

Or He didn't.

Death and suffering either first entered into the world after Adam committed the original sin in the Garden of Eden, just as it is reported in the Book of Romans...

Or it didn't.

The purpose of this report is to expose the lies of Darwinism. We are making our case strictly on how Darwinism stands up under the microscope of real science, and not hearsay or putting together strings of incredible assumptions as Secularists do. We will give numerous examples of their assumptions in this report.

Why Devolution is a most appropriate word.

God, by His own assessment, created a universe that was good. Planet Earth was perfect. Just imagine – a world without smog, no traffic jams, no armies, no wars, no politicians, no bad weather, no sickness or death, no food shortage, and perfect health and harmony.

Adam and Eve were the most amazing specimens of humans ever, with perfect bodies, and most likely able to access 100 percent of their brain capacity.

They contained the complete gene pool that later spawned the many different variations of people that live on Planet Earth today, including the giants the Bible reports descended from Adam. Some Bible scholars postulate that Adam and Eve very well could have been what we today would call giants. And I'm not talking runt-sized giants like the estimated 9-feet to 9-and-one-half-feet tall Goliath.

Yes, I said runt. One giant mentioned in the Old Testament was Og the King of Bashan. Here is how he is described in the Book of Deuteronomy 3:11:

> **For only Og king of Bashan remained of the remnant of giants; behold his bedstead was a bedstead of iron; is it not in Rabbath of the children of Ammon? nine cubits was the length thereof, and four cubits the breadth of it, after the cubit of a man.**

A cubit was the length from finger tip to elbow. If, say, the average cubic at that time, which was in the 14th century BC when Moses was leading the Israelites from Egypt to Israel, was 20 inches, the bed would be about 15 feet long by about 6-feet-eight inches wide. King Og very possibly could have been in the neighborhood of 14-feet tall, which is a tall neighborhood. (Bashan is in what is now Syria, east of the River Jordan and the Sea of Galilee.)

But this is not observable or testable today so we can only speculate.

One thing is certain: everything went downhill after Adam and Eve committed the original sin of disobedience to God. Consider these observable facts about the world we live in today: sickness and death are rampant, wars and famine plague many countries, natural disasters are commonplace, economies go through endless boom and bust cycles, no place seems to be safe and secure anymore, and it is said that few people today use more than a few percentage points of their brain capacity.

I could go on and on listing how things have gone haywire after Adam and Eve's expulsion from the Garden of Eden.

For example, look at the many examples of ancient buildings and structures that modern technology has been unable to duplicate. Modern archeology and engineering have not determined how those huge stones of the Egyptian pyramids were put in place. We have several theories but no one is certain, so to say that we are smarter than the original builders is insane. We have no idea or conception how intelligent and capable the Antediluvians were – the people who lived from the time of Adam until the global flood.

But we do have many examples that have survived in ancient writings, including the Bible. We can also study the buildings, monuments, and other incredible examples of pre-flood engineering abilities that were built with knowledge brought into the post-flood world by Noah and his family, such as the great pyramids in Egypt.

And these examples were of achievements AFTER the global flood which is described in such detail in the Book of Genesis. Keep in mind that Noah and his three sons carried with them into the post-flood world what knowledge they retained from their pre-flood experiences. That knowledge gradually became diluted during ensuing generations until most of it was replaced or lost.

Comparing our abilities to the abilities of our ancient ancestors would be, I think, like comparing gnats to eagles. We have devolved that much.

Certainly modern man has built upon the knowledge gleaned from Noah's family and on the past 4,000-plus years of man-made discoveries. However, to think or say that we are more intelligent than those ancient generations is foolhardy.

"Evolution is unproved and unprovable. We believe it only because the only alternative is special creation, and that is unthinkable."

Sir Arthur Keith (1886-1955), famous British evolutionary anthropologist and anatomist who wrote *Darwin revalued* (London; Watts, 1955)

FOUR

What is this war all about?

Whatever belief system you subscribe to – and everyone has a belief system – it is important to know what the war that is raging in the United States and around the world is all about.

Sure, this report centers on the Creation vs. Darwinism debate, but do you really know why this might be the single most important discussion topic in the world today?

And do you know why the first three chapters in the Book of Genesis are the real reasons why there even is such a debate?

Or why Secular Humanists invested so much effort during the previous century to gain control of the educational system and textbooks in U.S. public schools at all levels on up through college and post-graduate study programs?

The Great Heist

We documented how Secular Humanists have literally stolen a country's heritage from several generations of school children in **The Theft of America's Heritage**. They did this by rewriting history and no longer teaching that the United States of America was founded by predominantly Christian men on predominantly Christian principles.

Secular Humanists won landmark battles in 1962 and 1963 when prayer and then the teaching of Biblical Creation were banned from public schools. Those events represent a line of demarcation in U.S. history when the country turned its collective back on God, and swapped a long-standing Christian worldview for a Darwinian-based Secular worldview.

The Republic of the United States of America has been on an ever-increasing downward spiral ever since. [If you disagree, I suggest that you read **The Theft of America's Heritage** in which we document America's retreat from its Biblical-based foundations.] But it did not happen overnight. It took several decades of gradually deeper entrenchment into the country's educational system by the so-called progressive education espoused by John Dewey, a prominent atheist, although that fact is often not mentioned in biographical discussions of Dewey.

And even decades before Biblical Creation was erased from public school curricula, Darwinism had been taught as if it were real science in many textbooks. Generations of children have been taught that the Bible is no more than a book of made-up stories, that there was never a global flood, that there was no such thing as original sin – and that they ultimately have evolved into human form as the result of life somehow starting as *goo* in some primordial soup that spawned all life as we know and see it today.

This is taught as fact today; and as having happened totally without the involvement of the Biblical God.

Is it any wonder the U.S. has devolved from being a country whose main public school challenges went from kids chewing gum and talking out of turn to the violence of rape, aggravated assault and school shootings that we endure today?

Recipe for chaos

I propose that the following has contributed greatly to the Great Chaos that has settled on our nation:

No prayer in schools.
No teaching of Biblical Creation.
Teaching that humans evolved without a god being involved.
Teaching that humans are the most evolved animal; and that you are your own god.

It's truly a recipe for chaos. Look around. You can see it coming with the tremendous loss of respect each succeeding generation has for the nation's forefathers, history, morals and laws. You can see it in the falling attendance rates in churches, and in the compromised positions taught in far too many churches today.

What does this have to do with the first three chapters of Genesis?

It has everything to do with the first three chapters of Genesis. In these three short chapters the entire foundation for the Gospel of Jesus Christ is laid down. I use the acronym **COSt** to illustrate this foundation:

C is for Biblical Creation as recorded in the Bible;

O is for the Original sin committed by Adam and Eve in the Garden of Eden;

S is for the Separation from God that resulted from Adam and Eve's sin;

t in the shape of a cross stands for our **need of redemption** with our loving Creator.

The **COSt** is clearly spelled out in these foundational chapters of Genesis. To destroy the need for Jesus' redeeming sacrifice on the cross you simply have to destroy these Biblical foundations. You need only undermine peoples' faith in the **COSt**.

God's inspired Word declares it was man's sin which brought death into the world.

God did not create a world with death and suffering; otherwise, it would not be a perfect world, and God would surely not pronounce His Creation as *very good*—if it were full of death!

If God is God, He's infinite and eternal. Since we know that both good and evil exist, we must be able to explain them in light of this omniscient Creator. A being acts according to his nature. If God were both good and evil, He would, by necessity of His infinite and eternal nature, be infinitely good and infinitely evil. This concept would not work because God would then be in internal conflict/war and He would destroy Himself (which obviously has not happened). We can rule out that God is infinitely evil. Therefore, we must say that God is good.

So, since God must be infinitely good, His very nature demands that He would have created a world that is good...and that it is man's sin (the beginning of Moral Evil – which began when man stopped seeking God – and resulted in man's spiritual death, bringing about a state of meaninglessness, boredom, and guilt) that brought about the curse.

The existence of Natural Evil – toil, strife, old age, sickness and death – came about AFTER Moral Evil. It was instituted by God and is used by Him as a call back from our sin, as a reminder for us to stop and think and consider why we are here and what life is all about.

All Darwinian-based tenets are based on the concept that the earth is *billions of years* old, and that death and suffering was around for *millions of years* BEFORE man evolved.

Secular philosophy claims that it was *millions of years of death* that brought man into existence.

Were that true there would have been no original sin that brought death into the world. This brings into question any original sin having separated us from God. Without original sin and the resulting separation from our Creator there would be no need for God's redemptive plan for man, no need for Jesus' sacrifice on the cross. Secular Humanists, whose foundational beliefs are based upon Darwinian change being a fact, have not overlooked this. Here's a quote that proves my point:

"Destroy...original sin...If Jesus was not the redeemer who died for our sins, and this is what evolution means, then Christianity is nothing!"

Richard Bozarth made the above statement in *The Meaning of Evolution.*

I wholeheartedly agree with Bozarth's statement. If *millions of years leading to Darwinism* is true then God's Word simply is not.

It was apparent to Secular Humanists that to destroy America, they needed to undermine true Christianity by destroying the American people's faith in the God of Scripture. Realizing this, the attackers centered their crosshairs on the **COSt**.

I cannot overstress the importance that Christians fully understand the relationship between the **COSt** and the need for our Lord's redeeming death, burial and resurrection, because without this foundational understanding for why we need redemption with our Creator, our Christian testimony is at best compromised as mine was during the years I was a Theistic Evolutionist.

It is vital for Christians, or for anyone seeking the truth, to understand the importance of the early chapters of the Book of Genesis. Only by understanding the relationship between the **COSt** and the Gospel of the Lord Jesus can a person comprehend how an attack on Biblical Creation is an attack on the very Gospel of Jesus Christ.

Because this is so vital let's review the COSt.

The Biblical view is that God's perfect Creation was corrupted by man's Original sin which separated us from our loving Creator and required our redemption with Him. So there is the **COSt**: (C)reation; (O)riginal sin: (S)eparation; and the need for (t) Redemption at the cross.

The **COSt** is the very foundation for the Gospel of Jesus Christ.

Creation was perfect until
Original sin
Separated us from God, requiring
(t) redemption with God
The Gospel: Jesus Christ is our redeeming Savior.

This is the **COSt** and is why Jesus' sacrifice was required in order to reunite us with God. This is all laid out in the first three chapters of the Book of Genesis and Jesus Christ is our loving Redeemer, first promised to us in Genesis 3:15. Almost the entire rest of the Bible is the story of God's plan of redemption, leading up to when the Lord Jesus paid the total **COSt** for all of us who believe in Him, those of us who have faith in Him, as our redeeming Savior—whom He claims to be as found in the uncompromised Word of God.

Only by understanding the vital early chapters of the book of Genesis can Believers understand why Jesus Christ was required to come to earth, live a sinless life, suffer and die on a cross, be buried in a tomb, and rise again on the third day, His resurrection defeating death once and for all.

Praise God! What a loving Creator we have!

Jesus Christ, the Creator Himself, provided the **t**, the redemption, overcoming the separation from God that resulted due to Adam's original sin which corrupted God's perfect Creation. Jesus paid the death penalty for our sins, defeated death and fulfilled God's promise in Gen 3:15, reuniting with our Creator God all of those who put their faith in the Jesus Christ of Scripture who is our Creator, Redeemer and loving Savior...all of those who believe in Him.

John 3:16 For God so loved the world that he gave his only begotten son that whosoever believeth in him should not perish but have everlasting life.

By understanding the relationship between Biblical Creation and the Gospel of our Redeeming Savior, you can see that in order to destroy Christianity you need only to undermine people's faith in the **COSt.**

Then by understanding the vital relationship between the **COSt** and when death entered the world, either after Adam's original sin or *millions of years* before Adam existed, you can begin to comprehend why the age of the earth is such a vital subject in relationship to one's faith in our Biblical Savior.

We explain this in our third report: **371 Days That Scarred Our Planet**. Once you understand how the old-earth dating methods are based on an assumption of the gradual accumulation of the earth's strata layers, you will see that in order to destroy people's faith in the **COS†** you need only convince people that *millions of years of death* occurred before man came about.

This is why old-earth believers must deny the global flood. The global flood explains the rapid formation of the sedimentary layers which comprise the crust of the earth, washing away old-earth beliefs.

2 Peter 3:3-6

[**3**] Knowing this first, that there shall come in the last days scoffers, walking after their own lusts,

[**4**] And saying, Where is the promise of his coming? for since the fathers fell asleep, all things continue as they were from the beginning of the creation.

[**5**] For this they willingly are ignorant of, that by the word of God the heavens were of old, and the earth standing out of the water and in the water:

[**6**] Whereby the world that then was, being overflowed with water, perished:

Undermining peoples' faith in the **COS†** is what *billions of years of time leading to Darwinism* is really all about – even if 90% of the people who have been fooled into trusting in old-earth philosophies don't realize it.

Colossians 2:8 Beware lest any man spoil you through philosophy and vain deceit, after the tradition of men, after the rudiments of the world, and not after Christ.

We cover the age of the earth in **371 Days That Scarred Our Planet**.

Darwinism pollutes the Word of God.

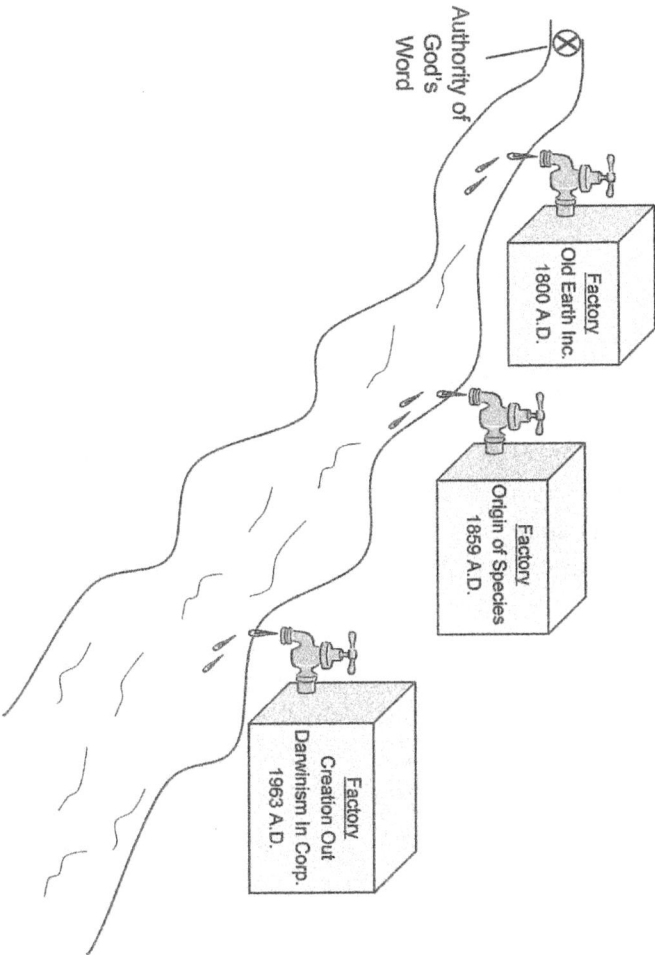

Authority of God's Word

Factory
Old Earth Inc.
1800 A.D.

Factory
Origin of Species
1859 A.D.

Factory
Creation Out
Darwinism In Corp.
1963 A.D.

Polluted Stream Branches:

Old Earth Gap Theory

Darwinism Theistic Evolution

1963 Creation & Prayer Out
"millions of years" &
Darwinism In

1966 Sexual Revolution

1966 Drug Culture

    Crime

    Family Breakdown

1970 Radical Movements –
Environmental
Animal Rights
Homosexual

1970-2008 Biblical
Compromise
Progressive Creation

2008 Post-Modernism – All
beliefs are equal

Today – 81% Christian Kids by
Age 20 leave the church

"The Creation-Evolution issue is not going to fade away as long as there is any measure of freedom of speech because freethinking people will not accept being told that there are some questions they are not allowed to ask and some answers that they are not allowed to question."

Often repeated by Russ Miller, founder of Creation, Evolution & Science Ministries Original source unknown

FIVE

Not ID.
But IBD.

There is Intelligent Design.
And there is Intelligent Biblical Design.

As I often tell Christians: "Every Christian believes in our Intelligent Biblical Designer, but not everyone who believes in Intelligent Design is a Christian."

Here's a true-life example that bears out this statement.

My co-author has a good friend who used to be a professing atheist even though he is a Jew. He has come to strongly believe that Intelligent Design is evident in the world we live in. He even believes there was a global flood as is recorded in Genesis; however, he sees no conflict with such a flood and *millions and billions of years*, which is the foundation of Darwinism. And he is adamant in stating, "I am not a Creationist!" He does not accept that Jesus Christ is the agent of creation as the Bible declares. Nor does he believe in the fact that it was man's original sin that separated us from our Creator. He certainly has not come to the point of acknowledging and accepting Jesus Christ as his personal Lord and Savior.

So not everyone who believes in Intelligent Design is a Christian.

Do I believe in Intelligent Design?

Absolutely!

Then why do I believe the emphasis is more correctly and properly focused in using the term Intelligent Biblical Design (IBD) vs. Intelligent Design (ID)?

Because I believe the Bible is true word for word and cover to cover. IBD acknowledges that our Biblical God is the Creator just as the first chapter of Genesis declares.

The Intelligent Design Movement (IDM)

Intelligent Design is an up-and-coming scientific field of study and there are some things that Christians need to understand about ID and the Intelligent Design Movement (IDM).

As I often remind people in my Power-Point presentations, the study of our origins has never been about the evidence because we all have the same evidence to study. The study of our world's origin is about the philosophical framework in which the evidence is interpreted. All such frameworks are based on axioms – starting beliefs. Darwinism is based on a starting assumption of Naturalism, that all observed processes have come about by naturally occurring phenomena. Meanwhile, Biblical Creationism is based on the non-compromised Word of God.

Both Biblical Creationism and Naturalism are religious beliefs, not demonstrable scientific facts. However, we can take things that can be scientifically observed today and compare those findings to the accounts of these two worldviews and see if either one meets its own predictions.

There are only three viable choices for why we exist:

1. The Biblical God created the world;

2. An unknown entity created the world; or

3. The world evolved naturally on its own.

Let's compare Intelligent Design, naturalistic Darwinism and Biblical Creationism.

The Intelligent Design Movement is based on the observable, intentional design in things. ID is a scientific theory which looks at the concept of "specified complexity" that is best explained as a result of an intelligent cause, and not as a product of natural random processes.

ID allows scientific findings to guide conclusions rather than having to fit discoveries into a pre-conceived worldview.

Naturalism, which is the foundational philosophy for Secular Humanism, is at its heart atheistic materialism – a religious philosophy based on the premise that "Matter is all there is." Naturalism is the belief that life itself and all systems, including the finely-tuned universe and complex biological systems in cells, evolved by purely naturalistic means over *billions of years*. Naturalism interprets all scientific findings through a materialistic worldview.

The New York Times warned that the Intelligent Design Movement "would...change the definition of science itself so that it is not restricted to the study of natural phenomena." This admits science is currently "restricted to the study of natural phenomena," to a materialistic worldview. Besides handcuffing scientific research, a huge problem with restricting research to the study of natural phenomena is that Naturalism is not a scientific theory, but rather it is a religious philosophy.

It is false to say that the Intelligent Design Movement is about getting Biblical Creation into schools. Not everyone in the IDM is a Creationist, much less a Biblical Creationist. Many IDM leaders believe in evolution, but not in atheistic, naturalistic evolution alone.

While every Creationist believes in an Intelligent Designer, not all Creationists are Biblical Creationists. While Muslims, Jews and Christian believers should agree with the Genesis account of creation, there are many creation beliefs which are not based on the Biblical account: for example, New Agers, Hindus, Deists, Agnostics, Progressive Creationists and Theistic Evolutionists are all fine with the IDM. Even the Dahlia Lama has spoken in favor of the teaching of ID. In fact, even those who claim that "aliens dropped us off" are fine with the IDM. Each group simply credits their deity with having provided the observed design.

The only religious belief threatened by ID is Secular Humanism whose foundation is materialistic, naturalistic Darwinism and, unfortunately, Secular Humanists own the textbooks, public education systems and scientific establishments.

Sadly, it may be accurate to expect that perhaps one day the Pledge of Allegiance will be amended to say that we are "One nation under an Intelligent Designer..."

It is propaganda.

To say the IDM is trying to get the Christian God into schools is propaganda put forth by those whose religious dogma is beginning to lose its stranglehold on the educational and scientific establishments.

From my Christian perspective, here is what I perceive as the IDM's strengths and weaknesses:

Perceived strengths of the IDM

The IDM has developed many materials which are very useful in strengthening arguments for Biblical Creation. These include books and DVDs on the complexity of both living and non-living systems. The IDM has also updated many issues and terms which Biblical Creationists employ. For instance, the terms Specified Complexity (signs of design) and Irreducible Complexity (too complicated to form slowly on its own) are better suited to scientific discussion.

Also, the IDM has successfully focused many scientists' attention on the inherent weaknesses of Naturalism and Darwinian Evolutionism, exposing both for the faulty religious beliefs that they are.

Furthermore, the IDM is attempting to free science from the grip of Darwinism so that researchers can return to the unbiased gathering of knowledge derived from the study of evidence. In other words, the IDM is helping to return science to being science as opposed to the proselytizer of Secular Humanism.

Forcing researchers to force-fit their findings to support Naturalistic Darwinism has undermined scientific research and education for the past fifty years. Even today scientists must toe the Darwinian line or face severe repercussions which include not getting hired, being fired, not having papers published, not receiving grant money to conduct research...in short, Humanism survives on blackmail and coercion. See the Ben Stein documentary *Expelled: No Intelligence Allowed* or the book *Slaughter of the Dissidents* by Dr. Jerry Bergman for examples of this religious coercion which has been dominating scientific education and research since 1963 and masquerading as if it were science for many years before that.

Past giants of science such as Sir Isaac Newton (Law of Gravity; Calculus), Louis Pasteur (Law of Biogenesis), Blaise Pascal (Hydrostatics; Barometer), Leonardo da Vinci (Hydraulics), and many, many others would have been shut out of today's scientific arena because each was a young-earth Biblical Creationist.

In fact, more than 80% of the branches of modern science were founded by young-earth Biblical Creationists who were searching for how our orderly God put things together.

However, *millions of years leading to Darwinism* has taken over and undermined educational freedom today. Competing hypotheses are banned from schools. Even teaching the weaknesses of Neo-Darwinism is banned in many school districts. Total censorship is the Secular Humanists' best friend because observable facts – real science – are their worst enemy.

Contrary to the thesis that the IDM is attempting to stifle science and free thought, the IDM is attempting to promote academic freedom.

Perceived weaknesses of the IDM

First, many leaders in the IDM are not Christians. In fact, several are openly hostile to Biblical Creation while others reject various Christian tenets such as original sin, the fall, and the global flood. I predict that, should IDM succeed and become an accepted branch of mainstream science (as it should), some of the strongest opponents to the one true Intelligent Biblical Designer, and certainly to Biblical Creation, will arise from inside IDM ranks. I absolutely expect this will take place should the IDM succeed.

While pointing out the failings of Naturalistic Darwinism, especially in biology, many leaders of the IDM support anti-Biblical findings in scientific fields whose interpretation of facts are based on the starting presupposition of Naturalism. This is especially evident with regard to geology. The great majority of IDM leaders (the term "leader" is a description of their public recognition as they are not appointed positions) believe in *millions of years* of death taking place before mankind came about.

Acceptance of the IDM will most likely lead to an increase in promoting non-Biblical forms of creation.

These increasing non-Biblical forms of creation will include the New Age movement and Hinduism, and even more compromise positions within the Christian Church. These compromised positions already include the Gap Theorists, Theistic Evolutionists, Progressive Creationists, Day-Age Theorists and more.

This is why I often like to ask Christians, "How many Jesuses did die on that cross so our sin could be forgiven?!"

In September 2004 the liberal *Christianity Today* publication called for Christians to stop attacking Darwinism, start supporting the Intelligent Design Movement, and begin attacks against young-earth Creationists. Well why in the world would they tell folks to attack those of us who believe God's Word? Because of the compromise of God's Word with various *millions of years* beliefs which are running rampant in today's Church.

It is a model of spiritual compromise that was evident during Jesus' time as described in the New Testament Book of John. John is describing the atmosphere of antagonism of the then-prevalent beliefs of those seeking to kill Jesus, and the truth of the Savior's teachings. Believers in Jesus began to hide their recognition of the truth He taught:

John 12:42-43 ...among the chief rulers also many believed on him; but...they did not confess him...For they loved the praise of men more than the praise of God.

We discuss this great compromise in detail in report No. 4: **The Submerging Church.** We document the vast damage done to the Christian Church by Secular beliefs being taught as if they were science in public education; and we discuss how the majority of Christian leaders have compromised on the foundational accounts found in the early chapters of Genesis. Even many professing Christians are blind to this great compromise in today's Church. In report No. 4 we pull no punches in exposing this degradation of God's Word.

Compromise has always been one of Satan's tools. Even the great apostle Peter was falling into compromise until Paul rebuked him, saving the church from legalism (read that account in Galations, chapter 2).

In a later chapter in this report we will discuss how ID, Darwinism, and Biblical Creationism stand up under the microscope of real science.

1 Thessalonians 5:21
Prove all things; hold fast that
which is good.

SIX

Amazing Examples of IBD

Imagine what our lives would be like without a lot of the things we take for granted. I'm talking about things such as cars, airplanes and space shuttles and antibiotics and improved healthcare, none of which would be around were it not for operational science.

A condensed scientific definition of operational science is the study of repeatable evidence. In simple terms it is finding out how things work in order to use that knowledge to make new inventions, products, and other things that bring vast improvements to our lifestyles.

We're talking real science, where things can be observed in labs and other settings and repeatedly tested to make sure the same results are achieved each time.

Darwinists like to lump operational science in the same pot with what is called origin science, or historical science, which does include the honest study of the past by such scientific endeavors as criminal pathology (finding clues as to how, why and who murdered a person, for example) and archaeology (finding clues about ancient structures and civilizations).

Where real science gets trampled upon is the incredible attempt to stretch origin science to fit the *millions of years leading to Darwinian evolution* belief required to hold up the platform of Secular Humanism. Speculation without any proof runs rampant. But even worse, speculation backed up by deliberate frauds has been running wild in school textbooks as well.

True origin scientists study evidence that is available and observable. They don't invent storylines to confirm a set belief, for example, in the notion that it took *millions and billions of years* for the universe to evolve into the state we see it today. Nobody has ever seen Darwinian-style evolution (one kind of animal or plant changing into another kind of animal or plant) take place. And, as we will show you in this report, no clear-cut fossil or other evidence ever has been found that supports Darwinism.

Real science, as I say in my Power-Point presentations, is a Darwinist's worst enemy, and a Christian's best friend.

One common thread throughout all of the advances in human-developed technology is that the successful medical researchers and design engineers did not sit around waiting for unplanned, random chance events to occur which would, hopefully, improve their well-designed technology. Instead, they had set goals and used massive amounts of human intelligence, built upon years and years of other people's research and study, in order to achieve their desired results.

With this in mind, it doesn't seem consistent or logical to believe that unplanned, random Darwinian processes are a viable explanation for the fearfully wonderful and extremely intricate design that is readily apparent in all people, plants and living creatures.

Not only that but the Second Law of Thermodynamics (which we discuss in a later chapter) concurs as well: things change, over time, from a state of complexity to simplicity. Nothing randomly goes from simple to complex on its own without intelligence and energy being involved in the process.

Let's look at some of the amazing examples of Intelligent Biblical Design (IBD) found in the things that sprout from the earth, swim in the seas, fly through the firmament or creep upon the earth – and let's look at ourselves as well.

> **Genesis 1:11 And God said, Let the earth bring forth grass, the herb yielding seed, and the fruit tree yielding fruit after his kind, whose seed is in itself, upon the earth: and it was so.**

After preparing the basic orb, water, atmosphere and dirt God made the food source for His original creation – plants.

Most people are aware that bees are needed to pollinate plants but the intricate design found in how bees and plants sometimes function together is yet another fingerprint of our Intelligent Biblical Designer found throughout His creation.

Consider again the Bucket Orchid as an example. The petal of a Bucket Orchid has a slimy surface which causes a bee, attracted by the aroma of nectar put out by the plant, to slip and fall into the orchid's "bucket." In the bottom of the bucket is a pool of liquid and the only way for the poor little bee to escape from the bucket is through a tunnel that goes from the side of the pool to the outside of the flower. There is even a step at the edge of the pool for the bee to climb up onto and go into the tunnel.

However, as the bee crawls through the tunnel and towards freedom, the walls of the orchid contract and capture it, holding the insect while the flower glues two pollen sacs to the creature's back.

After allowing time for the glue to dry, the flower then releases the bee so it can fly off on its search for nectar and if it lands on another Bucket Orchid the entire process is repeated. The bee slips and falls into the pool then crawls up the step and into the tunnel where the flower walls contract on it again. I am fairly sure that the little fellow is thinking, "Déjà vu! Wasn't I here before?"

However, this time two hooks come from the orchid and remove the two pollen sacs, completing the pollination process before releasing the insect to go about its business. Wow! Talk about awesome proof of our Intelligent Biblical Designer!

> **Genesis 1:20a And God said, Let the waters bring forth abundantly the moving creature that hath life...**

After taking a day to add the sun and moon, as well as the stars, God filled His sea with critters.

Dolphins are water-dwelling mammals that scream out **"Intelligent Biblical Designer!"** They can swim up to 25 miles per hour although, theoretically speaking, their muscles should only allow them to swim at about 10 miles per hour. Their ability to glide so fast through the world's ocean's is due to the way God designed and created them, with outer skin that is made of a material which becomes water-saturated and automatically adjusts to any movement the dolphin makes. This incredible system reduces the friction between the creature's skin and the surrounding water by as much as 60%, allowing the animal to attain speeds much faster than their muscles alone could generate.

Navies around the world and various scientists have been attempting to copy the IBD found in the dolphin's outer covering and apply simulations of that masterful system to ship hulls, submarines, torpedoes and more. Human successes in such endeavors are directly due to copying what has already been found in our Intelligent Biblical Designer's creation.

Genesis 1:21a And God created great whales, and every living creature that moveth, which the waters brought forth abundantly, after their kind...

OK, let's slow things down a bit and discuss a critter that has no heart or brain (no, it is not your old boyfriend or girlfriend). This incredible oceanic hunter contains no bones or eyes, breathes through the surface of its body and is made of about 98% water.

What is it?

Well, it's a jellyfish which is not really a fish at all.

There are two ways in which these invertebrate creatures move along in the water. Some swim, using a jet propulsion system whereby specially-designed muscles, built into the underside of their body, push water out of the hollow "bell" to push them along. Others attach themselves to material, such as kelp, and float along the ocean's currents with that item. Either way, their movement is greatly influenced by sea currents, tides and waves.

But do not let the non-threatening, lethargic appearance of these remarkable creatures fool you; jellyfish are extremely well-designed ocean predators. Created to be almost transparent, they are camouflaged wherever they go and they were created with tentacles that are used to catch their prey and flaps for devouring their food.

Without eyes to see, jellyfish rely on nerve cells to help them hunt and survive, reacting to food sources or danger. Their preferred prey include microscopic sea creatures, small fish and even other jellyfish. Using well-designed cells that contain a tiny harpoon that, once triggered by movement or touch, shoots a toxin into their victim, jellyfish disable, kill and then eat their prey. The potency of their toxin ranges among the varying kinds of jellyfish, causing reactions in humans that go from mild rashes to death.

The Artic Lion's Mane is the largest known jellyfish. Its body is about eight feet in diameter and it can weigh more than 500 pounds. With tentacles up to 200 feet in length this jellyfish can truly "reach out and touch someone!"

Genesis 1:20b ...and fowl that may fly above the earth in the open firmament of heaven.

I imagine that flight has fascinated mankind since the days of Adam. Only in the past hundred or so years have * modern humans begun to engage in controlled flight, and during that time we have made some tremendous advancements from airplanes to space shuttles to various types of space-probes.

* I used the word "modern" as evidence exists in ancient writings that include detailed drawings of ancient man's attempts to design crafts for flight; models of which have even been found in several ancient Egyptian burial chambers, including the one containing the remains of Tutankhamun, more commonly known as King Tut.

Often our technology was founded upon the design found in the fliers that God first created on the fifth day of His six days of creation.

There have been at least four types of created fliers in the world: birds, mammals, reptiles (extinct pterosaurs) and insects. Each is quite different from the other kind and the ability to fly reflects what is known as "irreducible complexity." This is the term that indicates all of the parts which allow for flight had to be there from the start or else the poor creature would have had to flop around on the ground as it evolved, vulnerable to easy predation – that is, they would have been removed by natural selection had things developed naturally over long ages of time in Darwinian fashion.

Think about this for a moment: none of the individual parts of a flier can fly without all of the other essential pieces. I heard someone say once that they were sitting on an airplane and began to think about how an airplane remains aloft. None of the various parts of the engine can fly; none of the pieces in the wing of the craft can fly; the wings themselves won't lift off of the ground. In fact, if you were to hold any piece of an airplane in your hands and let it go it would simply drop to the ground.

The fact is that all of the parts must be working together as a whole system just as intelligent human engineers designed them to function or the plane itself would drop from the sky.

The same can be said of living fliers. If you held a bird's feather, beak or tail in your hands and let it go it would simply drop to the ground.

Genesis 1:21 And God created...every winged fowl after his kind: and God saw that it was good.

Made on the fifth day of the creation week, birds scream out "IBD!" In fact, Darwin's finches, adapted through the loss of genetic material from the originally-created kinds of birds, also proclaim our **"Intelligent Biblical Designer."**

The aerodynamic wings of a finch work to keep the bird aloft by their flapping motion which forces air backward, over their wing feathers. The flapping action is gained by strong breast muscles which push the wings downward and a uniquely-designed pulley system which lifts the wings back up. The airflow over the wing feathers creates a hollow space above the wing which actually lifts the bird, allowing for flight to be possible.

Bird feathers are extremely complex. They are made of a hollow, thus lightweight, shaft and are crisscrossed with barbs which add strength while allowing the feathers to be flexible. The instructions to form a finch's feather are found in their DNA. This genetic information is different from genetic data which forms fingernails on people or scales on snakes – although each is formed from a variation of a substance known as keratin.

Bird bones are hollow and would break easily were it not for their internal cross-members which give bird skeletons a combination of great strength and light weight – a great design to allow for flight. This is also a great design for airplane wings and high-rise buildings, which is why intelligent human design engineers have incorporated such cross-members into these man-made structures.

Birds also have a very specifically-designed air-sac which lets air flow through it while supplying oxygen to their lungs. Their wishbone provides strength and flexibility while protecting their chests; and even a bird's beak is designed of a light horn material.

All of the bird's various parts work together to allow for flight. The muscles that operate the finch's wings and tail feathers work in perfect conjunction with the flier's central nervous system to control flight. All of the movements needed for flight are guided and coordinated by its on-board avionics computer, the bird's brain. Being called a "bird-brain" may not be such a bad thing after all.

Genesis 1:24 And God said, Let the earth bring forth the living creature after his kind, cattle, and creeping thing, and beast of the earth after his kind: and it was so.

Let me be honest with you – and this won't likely come as a shock – but I don't know everything. There, I said it; boy is the pressure off! Even the smartest human in the world may know very little in relation to all there is to know. What I want to own up to not knowing in this case is when critters such as bats and flying insects were made. I suspect they were made on day six but other well-known Creationists believe Scripture testifies to them having been created on the fifth day. I really am not sure.

Whether made on day five or on day six, when it comes to hunting and capturing prey in the post-curse world, few animals can claim to have such an intelligently and Biblically-designed system as the insect-eating, flying mammal known as a bat. And despite stories to the contrary, bats do not lay their eggs in people's hair causing them to go crazy.

Bats rely on "bat sonar" as it is often referred to which is a highly designed echo-location system which allows them to hunt insects during the nighttime. Insect-devouring bats send out a high-pitched sound, a sonar pulse, at from 30,000 to 100,000 hertz (cycles per second). The flying mammal then listens for the echoes bouncing off of nearby objects. By identifying these echoes a bat is able to avoid obstacles, identify insects, and then track and capture its prey. Some bats can catch as many as 500 mosquitoes per hour.

Fortunately for us the upper limit for human hearing is 20,000 hertz so their sounds don't drive us batty.

The "bat sound" is an intense, high-frequency ultrasound which echoes back from objects to the bat whose large ears divert the ultrasounds to its inner ear where a muscle holds the hearing bones while the pulse is emitted and relaxes in time to receive echoes from the previous pulse. This is totally incredible to ponder as all of this happens in a few thousandths of a second and allows the bat to distinguish its prey, other objects and even the ultrasounds emitted from other bats – all the while during full flight. Really incredible. In fact, absolutely amazing!

As usual, the supposedly oldest bat (by secular accounts this is the one found in the lowest strata layer) is just as highly designed as any of the bats found alive today and, "There are no known intermediate stages between bats and insectivores." Hickman, Larson & Roberts; *Evolution of the vertebrates*; page 333; 2001.

How well designed is "bat sonar?" Well, man-made sonar can distinguish ultrasound echoes 6 millionths of a second apart, while a bat's can identify echoes just 2 millionths of a second apart. That's what I call the fingerprint of God upon His creation.

And then there are those amazing butterflies and those rascally bees. Again, whether from day five or day six, they both certainly scream out **"Intelligent Biblical Designer!"**

Butterflies are incredible creatures that contain encoded hereditary genetic information which first develops a crawling caterpillar, then later the pupa, and finally a majestic flying butterfly. This genetic information is so compactly stored that the assembly instructions to form 10 billion butterflies could be held in a container the size of an aspirin!

A caterpillar has strong jaws for chewing up its food source, the leaves of plants. The caterpillar's entire digestive system is specifically designed to digest leaves, often just the leaf from one specific kind of plant.

Once the caterpillar is fully grown it spins a silken anchor onto a leaf and goes into its pupa phase in which it has neither legs nor a head. It is in this pupa stage that a miraculous change, or metamorphosis, occurs. The caterpillar's organs, with the exception of its central nervous system, dissolve into single cells from which new and different organs develop to form a completely different creature, a butterfly.

The butterfly emerges with the ability to fly and with a long retractable sucker for feeding on flower nectar. It also has long jointed legs to cling to flowering blossoms, compound eyes, and for many butterflies, the navigation ability to migrate thousands of miles – to lay its eggs on the leaves of the same kind of plant that it fed upon when it was a caterpillar. All of its characteristics and abilities were encoded from the start in its highly-designed, inherited genetic data, including its instincts for finding food and surviving in each of its three well-designed and very different stages. Awesome proof of IBD!

We discussed earlier how bees interact with the amazing design found in the Bucket Orchid; however, honeybees themselves are another of the wonderful examples of design found in God's creation. Honeybees are team players that construct their well-designed honeycombs in a nest or hive. To accomplish this bees live in colonies with up to seventy-five thousand members, in which team members have specific, and often differing, responsibilities.

Bee colonies are made up of a queen bee, female worker bees and male drones.

There is only one queen bee and the drones' one purpose in life is to mate with her. In fact, the male dies after completing this task. While the queen bee may live up to eight years, the female worker bees only live about twenty or so days. The worker bees perform the hive-building, with some specializing in honeycomb construction, others in honey-making and still others in guarding the nest against enemies. Younger worker bees maintain the hive's temperature by fanning their wings to cool the hive down or by huddling together to warm the hive up. In their efforts to obtain resources, honeybees transfer pollen from flower to flower completing the pollination process for plants such as the Bucket Orchid described earlier in this chapter.

One of the most incredible design features found with bees is their method of communicating with each other, which is known as the "honeybee dance." The various movements of a "scout" bee tell other worker bees about discovered food sources, its direction and its distance from the hive, and more.

Every movement by the scout bee conveys information to the other bees. For instance, if the food source is within a hundred yards of the hive the worker bee will turn about three-fourths of an inch in a tight circular motion, and then change directions. If the food is further away, the worker makes a circular pattern and then moves across the circle. The more times she goes through the circuit, the further it is from the hive and the direction she moves across the circle tells the others which direction it is from the hive, relative to the sun. What a truly incredible communication system they have.

The precise movements were required from the start and scream out, **"Intelligent Biblical Designer!"**

Genesis 1: 24 ...and beast of the earth after his kind: and it was so.

I was speaking at some churches which were spread throughout the Alaskan bush country a few years ago. These are areas only accessible by small plane or boat. A fellow told me that he needed to get his freezer filled with his family's winter supply of meat and he asked me if I would like to go moose hunting with him for a day. I said sure and began inquiring about what I needed for mosquito and bear protection.

A readily available spray can took care of the smaller pests. However, every time I asked someone for their opinion about "bear pepper spray" all responded: "Sure you should carry some pepper spray; the bears love their food spicy."

Before sunrise the next day we loaded into a 16-foot aluminum boat and headed upriver 16 miles, disembarking on a wide sandbar which was covered with bear tracks – BIG bear tracks. I am referring to tracks twenty-two inches long, 14 inches wide and sunken three inches into the sandbar. The type of bruins that love their food spicy.

Well, we hiked around all day and never saw a moose or a bear and when we returned to the sandbar that held both the boat and the bear tracks late that afternoon, the engine wouldn't start. It evidently had a history of conking out so the local fellow said, "I guess we will just have to sleep on this sandbar tonight."

I said, "I don't think so." I looked around and pulled a couple of ten-foot-long alder poles from a nearby beaver dam and we poled our way downriver, arriving back in town about midnight.

Beavers and bears are both animals that display the fingerprint of their Intelligent Biblical Designer upon them.

Beavers are furry mammals that spend a lot of time underwater and their ability to remain submerged for long periods is due to their incredible design features. For instance, they were made with a heartbeat which slows down when they go under water. The slowed rate is coupled with their large liver and oversized lungs which, together, let them store large volumes of oxygenated blood which is then supplied to their brain, allowing beavers to stay submerged for extended periods. In fact, beavers have been known to swim a half-mile while submerged

Other beaver traits which reveal our Intelligent Biblical Designer include that the beaver's eyelids are transparent and act like the facemask of a scuba diver, allowing a submerged beaver to close its eyelids to protect its eyes while still retaining its ability to see. Also, when it goes underwater, specially-designed valves in the beaver's ears and nose close so that water doesn't flow in. These valves then reopen once the critter surfaces. Beavers even have special fur-lined mouth flaps between their front incisors and rear molars which seal off their mouths and allow them to eat underwater without having water rush down their throats, drowning them.

Webbed feet for swimming; a paddle-shaped tail that allows it to steer while dragging logs and branches through the water; a self-oiling glandular system which prevents water from getting to the creature's skin; and its self-sharpening incisors all scream out, **"Intelligent Biblical Designer!"**

The tremendous engineering skills programmed into the beaver's brain also screams out, **"IBD!"** Beaver lodges are an engineering marvel, with the only access coming from underwater through multiple "doorways" for entrance and escape routes. Also, they usually build spillways in their dams to keep them from bursting...so long as someone doesn't come along and remove poles from the dam.

> **1 Samuel 17:37 David said moreover, The LORD that delivered me out of the paw of the lion, and out of the paw of the bear...**

Bears come in a wide variety of adapted kinds, from the black bear to the polar bear and several kinds in between. Although God could have brought many beK-Arinds to Noah, these all likely adapted from two bears that emerged from the ark, accumulating various genetic losses through adaptations and mutations (Gene

Depletion) to arrive at the bruins found around the globe today, many of which can still inter-breed.

The large brown bears that left their tracks on the sandbar described previously are awesome and powerful creatures that can store up to 400 pounds of fat by late fall and this fat acts as their food source while they den up through the long Alaskan winters. Their ability to store and use their fat while in a state of semi-hibernation may shed some light on how they, and perhaps other animals, spent their year on the ark of Noah.

The largest brown bears found in Alaska can weigh as much as 2,000 pounds and brown bears can run up to thirty-five-miles per hour – much faster than an Olympic sprinter.

This fact reminds me of the old Alaskan joke about the bear hunter and his hunting guide. While preparing to set off from camp on their bear hunt the hunter asked the guide whether he should wear his heavy hiking boots to protect against the Alaskan cold or some lightweight hikers in case he should need to run from a hungry bear. The guide answered that he should wear the heavy boots as it is impossible to outrun a bear. As the hunter was lacing up his heavy boots he noticed that his guide was tying on a pair of light hiking shoes and asked him why the guide wasn't following his own advice. The guide informed him that he wanted to be able to run swiftly should an angry bruin take after them to which the hunter stated, "You told me a person could not possibly outrun a bear!" The guide retorted, "That is true, but I wouldn't need to outrun the bear, I would only have to be able to outrun you!"

Ouch. Spicy food with heavy boots.

The brown bear's six-inch-long claws are perfect for digging and they feed primarily on roots, plants, berries and fish. However, they do occasionally catch deer and moose which become a part of their diet as well.

Their powerful jaws were well-designed for crushing plant material but in our cursed, post-flood world they can crush up moose bones and spicy food as well. I kept this thought in the forefront of my mind while looking around for a moose for my friend's family to munch.

Psalm 139:14 I will praise thee; for I am fearfully and wonderfully made: marvellous are thy works; and that my soul knoweth right well.

And what about us?

Genesis 1:26a And God said, Let us make man in our image, after our likeness...

Well, God saved the best for mankind. Take our personal computer for example.

Our brain is one of the best gifts put in us by our Intelligent Biblical Designer. The three or so pounds of human grey matter is in reality a living computer which is far more complex than anything ever put together with the human intelligence it contains.

Our brain sends out billions of pieces of information to trillions of nerves throughout our body via the central nervous system, controlling every move we make. The human brain distinguishes sounds received through our ears, molecules breathed in through our noses, things felt upon our skin and light waves taken in through our eyes and turns them into information we can use. From the beat of your heart to the twitch of your eye, your brain controls every action and reaction.

Avowed atheist and Harvard University professor, the late Stephen J. Gould wrote:

"...the pageant of evolution (is) a staggeringly improbable series of events...the chance becomes vanishingly small that anything like human intelligence would grace the replay."

All the same he chose to trust in such a staggering improbable series.

Psalms 94:9a He that planted the ear, shall he not hear?

The human ear contains a small musical instrument that was designed with three tiny bones that, from birth, never change in size. These are called the ossicles and they are the smallest bones in your body. Without this mature-at-birth design we wouldn't be able to hear until we were adults and the bones were fully grown.

The human ear is capable of hearing high-pitched sounds at up to 20,000 hertz to as low as 20 cycles per second.

The outer ear is cup-shaped, well designed to capture sound waves and divert them into the ear canal which funnels them to the eardrum. The miniscule vibrations of the eardrum are then passed on to the three tiny bones which are located in the middle ear. These bones act to compress the vibrations and pass them to the cochlea which is found in the inner ear.

The cochlea is shaped like a tiny snail shell and is made with outer and inner rows of "hairs." Microscopic vents open and close at up to 20,000 times per second to allow charged ions to reach the tips of the hairs. In doing so the cochlea changes the vibrations it receives into electrical signals which are transmitted to, and understood by, your brain. Fantastic IBD.

The human ear employs bone, fluid and air to transport sound waves; it contains a self-cleaning and lubrication system; the ear is completely integrated, exhibits irreducible complexity and screams out **"Intelligent Biblical Designer!"**

Psalms 94:9b He that formed the eye, shall he not see?

As you read these words you are actually seeing and translating them with your brain. However, your brain requires another incredibly well-designed organ to carry the visual images to it. This organ is the human eye.

With an estimated 1.25 million nerve connections in about a one-square-inch area, your eyes are an amazing system which communicate back and forth with your brain to allow you to catch a ball, shoot an arrow, or read and understand the words on this page.

Charles Darwin himself realized that it seemed incredible that naturalistic evolutionary processes could explain human vision and at that time we knew little of the complexity involved in sight compared to what we know today. Darwin wrote:

> **"To suppose that the eye with all its inimitable contrivances for adjusting the focus to different distances, for admitting different amounts of light, and for the correction of spherical and chromatic aberration, could have been formed by natural selection, seems, I freely confess, absurd in the highest degree."** (*The Origin of Species*, J.M. Dent & Sons Ltd, London, 1971, page 167.)

All the same Darwin chose to see the eye through what he referred to as the "highest degree of absurdity" – Darwinian-style evolution.

Our brain is the control center for the human eye, including its ability to focus. This is not something we have to think about doing as our brain focuses our eyes automatically when functioning together with the eye as both were designed to do. Today's cameras operate on the same basic principles as our eyes do.

The human eye even came with its own lubrication system – tears – which lubricate both your eyeball and your eyelid, preventing the dehydration of the mucous membranes. You were even designed with a gland in your eyelid that produces an oily substance which reduces the evaporation rate of your tears. This lubrication system is one of the many fingerprints of Intelligent Biblical Design in the human body as without this anointing system in place from the start, severe pain leading to blindness would have quickly set in.

Another of God's fingerprints is that the human eye also produces its own antibacterial and antiviral agent, a protein known as lysozyme. This is the major source of protection for your eyes from outside invaders and without this specific protein having been there from the very start, severe pain leading to eye infections would quickly lead to blindness and death.

John Stevens, professor of Biomedical Engineering, stated that it would take a Cray supercomputer 100 years "to simulate what takes place in your eye many times each second." (*Byte*, April 1985).

There have been at least five different kinds of eyes: humans; anthropoids, squids, vertebrates and trilobites. This means that to accept Darwinism, you must believe that eyes evolved not just once, but at least five different times completely independent from each other. And trilobites had a double lens with up to 15,000 lens surfaces – we humans have a single lens design.

Yet trilobites are found in some of the lowest strata layers – supposedly their eyes were one of the first things to have evolved on their own – ludicrous! Eyes of any sort, whether trilobite or human, are great proof of our Intelligent Biblical Creator.

Still, *Science News,* blinded by the bias of naturalistic beliefs, reported that the eye of a trilobite, highly complex and found in the lower strata layers, had "the most sophisticated eye lenses ever produced by nature." (*Science News*; 2-2-74; page 72)

> **Deuteronomy 5:29 O that there were such an heart in them, that they would fear me, and keep all my commandments always, that it might be well with them, and with their children for ever!**

Our heart actually is a muscular pump that pushes an average of one-and-a-half gallons of blood per minute through thousands of miles of blood vessels, providing life-giving oxygen and food to every part of your body.

> **Romans 10:9 That if thou shalt confess with thy mouth the Lord Jesus, and shalt believe in thine heart that God hath raised him from the dead, thou shalt be saved.**

In our hearts we know that we were made by our Intelligent Biblical Designer.

Romans 10:10 For with the heart man believeth unto righteousness; and with the mouth confession is made unto salvation.

Consider this warning from noted astronomer, cosmologist and mathematician Sir Fred Hoyle:

"Be suspicious of a theory if more and more hypotheses are needed to support it as new facts become available...this is exactly what has happened to Darwin's theory."

SEVEN

Seven Seconds Flat!

This is the chapter where I demonstrate that Darwinism can be refuted **– in Seven Seconds Flat!**

This is not a misprint.

It can be done.

No smoke and mirrors involved.

Just scientifically-proven facts.

I repeat:

Darwinism can actually be scientifically refuted in SEVEN SECONDS FLAT—or faster.   I will time myself just before we finish writing this portion of the chapter and tell you how I did.

Three facts are key to this demonstration.

Fact Number One:

The **Code Barrier**, best referred to as the DNA Code Barrier, is a scientific principle that one kind of plant or animal only has the genetic information in its gene pool to produce its own kind. Take a cow as a simple example.   While there may exist the genetic data to produce a wide variety of adaptations within the cow's particular DNA, the simple fact is that cattle only possess the genetic ability to produce other cattle.

Plants and animals can only bring forth after their kind, just as we are told ten times in the Book of Genesis. This fact is known as the DNA Code Barrier and is a major problem for Darwinian-style evolution. The Darwinist must try to explain some way for cows to produce non-cows, and this would only be feasible if there were a method for nature to add massive amounts of new and beneficial genetic information to an existing gene pool.

However, as of this writing, real science knows of no way for nature to add appreciable amounts of new and beneficial genetic data to one kind of a plant or animal's DNA. The DNA Code Barrier is a huge roadblock for Darwinism.

Fact Number Two:

Gene Depletion, also referred to as Genetic Entropy, is the scientific principal that all adaptations and/or mutations are the result of the sorting or the loss of the parents' genetic information. In other words, adaptational variations, as well as mutational changes, are caused by the recombination or loss of the beginning genetic data which was inherited from the parents and not by the gain of new and beneficial genetic data as Neo-Darwinism falsely teaches.

Managing gene depletion is how ranchers breed cattle to best match their end-use goals. Breeders can get meatier cattle or better milk producers by breeding out traits that they do not wish their cows to have. This is done through the loss of genetic information, not by the gain of new data in the cow's DNA. This is why the loss of genetic data is referred to as the scientific principle of Gene Depletion.

And managing this Gene Depletion, although it may produce meatier cows, will NEVER produce apes or humans.

Gene Depletion is yet another major roadblock for Darwinism, whose proponents keep insisting that humans, plants and animals are evolving better and better by the accumulation of new and beneficial genetic information being added to their gene pools, contrary to a mountain of evidence that shows that humans, plants and animals are devolving through Gene Depletion.

Fact Number Three:

Natural Selection is the process scientists observe during which, in the free competition for resources, the weaker members of a species tend to be eliminated, unable to compete with the stronger of their kind.

Now let's put these three scientific principles together to reveal why no one has ever been able to discover any viable evidence of Darwinism having taken place.

Start your timer:

"The DNA Code Barrier plus Gene Depletion plus Natural Selection makes Darwinian-style change scientifically impossible."

Stop your timer.

I got 6.4 seconds.   That is how long it just took me to scientifically refute Darwinism, the foundation for Secular Humanism.   And I was not talking fast like I sometimes do.

Background and discussion

In 2002 I began teaching people how to destroy Darwinism in seven seconds flat. I believe it is well worth taking a further look at these principles because once you understand the meaning of the three terms in this demonstration you will never be fooled again by Darwinian claims. You will also be able to quickly refute the religious belief of Darwinian-style evolution. Not only will you be able to scientifically destroy Darwinism (or Neo-Darwinism, which is what is actually taught today), but you will be able to defeat any Darwinist anywhere in the world, from Oxford to Harvard to your local high school in a debate. It is little wonder why Darwinists refuse to debate the topic anymore.

The scientific principles involved here include both the Code Barrier, which we introduced in Fact Number One, and Gene Depletion, introduced as Fact Number Two.

I refer to the Code Barrier as the DNA Code Barrier as it better defines the principle. The DNA Code Barrier basically states that the DNA of one kind of plant or animal only contains the genetic instructions to produce that particular kind of being. Note that there can exist in the gene pool of any particular kind the genetic variation to produce a wide range within that same kind. For instance, a pair of dogs may possess the gene pool to produce short-haired pups or long-haired pups, and they may birth black dogs or yellow dogs. These are simply variations, micro-adaptations within that same kind – dogs in this case.

Biblically-correct micro-adaptations, which are a scientific fact, always produce the same KIND of plant or animal. Millions of examples of adaptations within the same kind can be observed.

The fact remains that one kind of plant or animal only contains the genetic instructions to produce that particular kind of plant or animal, with some variation within that kind – the DNA Code Barrier.

Ranchers and Dairy farmers breed cattle for various end uses. Some are developed, through the intelligent selection of genetic traits found in the parents, that produce more meat while other cows are bred which produce more milk. But cattle will only produce cattle. Never will a cow produce a non-cow, such as a whale or a whitetail deer. That would be an example of Darwinian-style change – and never has such change been observed to have occurred. The DNA Code Barrier is one of the reasons why Darwinism is scientifically impossible.

To overcome the fact that one kind, like a cow, only has the genetic information to produce other cattle, Darwinism requires that massive amounts of new and beneficial genetic data be naturally added to an existing gene pool. Keep this in mind as we take another look at Gene Depletion.

Fact number two: Micro-adaptations are the result of the sorting or the loss of the parents' pre-existing genetic information. This is the scientific principle called Gene Depletion. In other words, adaptations lead to weaker and weaker gene pools. This is exactly the opposite of what Darwinism would require and is another reason why Darwinism has never been observed to have taken place.

With the religious belief of Darwinism in need of an observable method for nature to be able to add massive amounts of new and beneficial genetic information to an existing gene pool let us move on to fact number three: As of this writing, scientific knowledge, which is based upon the study of testable, observable evidence, knows of no way for nature to add appreciable amounts of new and beneficial genetic information to an existing gene pool.

After millions of observations, every single mutation and adaptation has been the result of the recombination or LOSS of the parents' genetic data – Gene Depletion. Once again you can see that what is actually observed is exactly the opposite of what Darwinism requires. The scientific principle of Gene Depletion is another reason why Darwinism has never been observed to have taken place. Not even once.

It is a major goal.

One of my goals is to present our life-changing information at a level that will be understood by folks aged nine to ninety-nine. In fact, being able to effectively communicate to a general audience is a weakness of many scientists but, being a general manager steeped in a business background, being able to use my God-given abilities to communicate with a wide range of people is one of my strong points – and I do not have a lot of them!

After presenting our **50 Facts versus Darwinism in the Textbooks** at a church one Sunday evening, a local college biology professor who heard of the seminar and attended raised his hand during the Q&A session that followed.

He promptly announced his high level of education and how he could show millions of examples of Darwinian change. I stated that he could not do so and he responded that he could give a thousand examples right on the spot. I did not wish to embarrass him but he had me in a position where I had to let everyone see who was correct, the professor or the Bible.

I said that he did not need to show a thousand examples of Darwinian change. In fact, he did not even need to produce ten because if he could produce just two examples, then I would conclude the Bible is not true and that would be that.

He promptly announced that butterflies that ate flower pollen had evolved into banana-eating butterflies. Before I could respond a nine-year-old girl stood up and said, "Professor, that is an example of a micro-adaptation, not Darwinian evolution." Although it was an embarrassing moment for the professor, God certainly used it to reveal the misguided and foolish faith adhered to by secularist educators to everyone in the audience.

200 branches of modern science

Non-scientists should note that there are about 200 branches of modern science. Because of this, a scientist who works in one of these branches is thoroughly trained in one-half of one percent of science. A scientist should be an expert in his small piece of the scientific pie, but he is not an expert in the other 99.5 percent of the scientific fields.

I have discovered that most scientists agree that their particular area of science does not hold much data to support Darwinism; however, I have also observed that scientists tend to believe that the other 199 or so branches have *the evidence*. This belief goes around the circle of the various branches of research with each area assuming the OTHER branches have the Darwinian-supporting evidences when in fact none of the branches of science contain any hard facts providing proof of Darwinism ever having occurred.

It is also a simple fact that scientists are people just like everyone else. All of us can make perfectly correct observations yet still come to the wrong conclusions. This is what happened to Charles Darwin when he sailed on the *HMS Beagle* to the Galapagos Islands. While there, Darwin made a brilliant observation. He counted thirteen varieties of finches on the islands (there are 14). Darwin observed thick-billed finches and thin-billed finches. He counted black and yellow finches and so on. He made a great observation, no doubt about it. But then Darwin jumped to the miraculously erroneous conclusion that somehow that proved that, "It is a truly wonderful fact…that all animals and all plants throughout all time and space should be related to each other…"

Great observation…horrendously wrong conclusion!

Darwin had not observed an example of Darwinism in action. Instead, he had seen evidence of variations within kinds – a Biblically-correct principle of micro-adaptations in action, but still kinds bringing forth after their kind through the sorting or loss of genetic information – Gene Depeletion. Then he wrote his misguided theory and the world soon jumped on board.

Where's the proof?

Secular-believing scientists and educators have been trying ever since 1859 to find proof in support of Darwin's theory but, with the exception of their biased interpretations of the evidence, they have not been able to support Darwin's miraculously-erroneous conclusion.

In fact, Darwin's *theory* was abandoned long ago. Today kids are misled into believing in another erroneous belief; this anti-scientific teaching is referred to as Neo-Darwinism.

For several decades schoolbooks have been teaching open-minded youth that, "Change was very slow because it relied solely on mutations."

I received this email from one misled man:

> **I was Roman Catholic for 40 of my 52 years and studied the Bible and its history intensely. We now have abundant, observational evidence of evolution (Neo-Darwinian) in action from the lab. This evidence includes fruit flies subjected to mutations. James**

So mutations are what add the required massive amounts of new and beneficial genetic information to existing gene pools to lead to Darwinian-style change? Again, let us prove all things and hold fast to that which is good. Neo-Darwinism is based on three false assumptions.

First:

Humanists teach that some mutations are caused by the addition of new and beneficial genetic information.

Second:

The secular crowd claims that it is Natural Selection that causes the supposedly improved mutant to take over the population.

Third:

Lastly we arrive at the third of Neo-Darwinian's false conclusions – in fact this is the magic ingredient for all forms of Darwinism: **long ages of time.** Darwinists require time that is beyond human comprehension, *billions and billions of years* of never-observed time leading to the formation of the earth, followed by *hundreds of millions of years* of non-testable time.

Neo-Darwinism teaches that somehow, given enough time, non-living matter produced a *simple* little single-celled creature magically overcoming the scientific Law of Biogenesis (a proven law stating that living matter only comes from other living matter), and overcoming all mathematical probability – and opened the door for all life forms to come about through Neo-Darwinian evolution without God being involved. Ridding the world of God is of course what *millions of years leading to Darwinism*, the foundational philosophy for Secular Humanism, is really all about when you clear away all of the debris. This allows people to believe that they are the most evolved of all creatures and are in essence their own god.

The Astrobiology professor at Northern Arizona University who began an accredited course attacking Biblical Creation during the spring semester of 2008, and specifically targeting Creation, Evolution & Science Ministries' teachings, told me that his dad was a Bible college professor. This misled NAU teacher said that he had wanted to be a Christian himself. However, he claimed that when he went to college and saw the preponderance of evidence for an old earth he became convinced that the Bible was simply not true.

Old-earth beliefs are the real culprit.

Most people do not realize that old-earth beliefs are the real culprit, and not Darwinism. Former Harvard professor and Nobel Laureate George Wald said it quite well when he stated:

> **"Time is in fact the hero of the plot...the impossible becomes possible; time itself performs the miracles."**

Millions of years beliefs undermine the **COS†** by placing death before man. This eliminates original sin having brought death into the world. Remember that God's Word tells us that this original sin is what separated us from God and required that we be reunited, or redeemed, with God. By undermining people's faith in the authority of God's Word about the **COS†**, old-earth beliefs undermine the need for redemption. They undermine the need for Jesus' earthly life, death and resurrection.

We thoroughly deal with the religious belief in *millions and billions of years* of time in report No. 3: **371 Days That Scarred Our Planet**. For now, we will stick with the testable and observable biological facts with regard to Darwinism.

The bottom line is not friendly to Neo-Darwinists.

Here is the bottom line problem for Neo-Darwinism: Following literally millions of observed mutations, whether scientifically-induced or naturally occurring, all such mistakes are caused by the sorting or the loss of the parents' genetic data.

That is correct.

Never has a mutation been observed which has added new and beneficial genetic information to the pre-existing gene pool which was inherited from the parents of the mutant. Remember the scientific principle of Gene Depletion? Gene Depletion applies to mutations just as it applies to adaptations.

This makes the mutant the most likely to be removed by natural selection.

So yet again real science reveals that what is actually observed is exactly the opposite of what is taught in the secular-owned textbooks. Neo-Darwinism requires the massive addition of new and beneficial genetic data; however, mutations are caused by Gene Depletion which is the sorting or loss of genetic information. The scientific principle of Gene Depletion is a major reason why Neo-Darwinism has never been observed to have taken place. Not even once.

In fact, Dr. H. Muller, a Nobel prize-winning mutation expert stated that:

"...good ones [mutations] are so rare that we can consider them all bad."

And he did not supply an example of a mutation which added new and beneficial genetic data to an existing gene pool.

So yet again, real science, knowledge which has been derived from the repeatable observations of testable evidence, is a Christian's best friend and a proven enemy to Secular Humanism.

In real science Darwinism would have been thrown into the dust bin of erroneous conclusions long ago. However, Darwinism is not science. The mistaken hypothesis of *goo-to-you* evolution is a religious dogma that has taken over the educational and scientific establishments, greatly undermining both endeavors.

Another of the multitude of secular lies designed to get people to not hear both sides of the issue is that a person who doesn't believe in Darwinism is "attacking science!" Nothing could be further from the truth. I am the one who is actually standing up for and employing real science.

Trying to rid the world of this already-refuted belief has caused many brave and honest scientists to lose their careers. As of this writing, anyone working in the educational or research fields who dares not toe the secular line of Darwinian belief will be severely punished. Loss of job, blackballed from other work, papers not getting published, grant money ending…the list of repercussions against honest appraisal of the Darwinian myth usually destroys an honest scientist's career instantly. Again, see the Ben Stein documentary **Expelled: No Intelligence Allowed** or the book **Slaughter of the Dissidents** by Dr. Jerry Bergman to get a perspective into the horrors endured by truth-seeking scientists and educators.

Since Darwinism is not observed to have taken place, and since it is the foundational philosophy of Secular Humanism, rather than casting it aside and moving forward with scientific research, more and more excuses, hypotheses and out-and-out frauds are employed to prop up the dead horse of Darwinian thought. In following chapters of this report we will examine many of the refutable claims that are taught in textbooks today for the purpose of propping up Darwinism.

Meanwhile, consider again this warning from noted astronomer, cosmologist and mathematician Sir Fred Hoyle:

> **"Be suspicious of a theory if more and more hypotheses are needed to support it as new facts become available…this is exactly what has happened to Darwin's theory."**

Renowned evolutionist Pierre-Paul Grasse wrote in *Evolution of Living Organisms* that:

"Mutations do not produce any kind of evolution."

Famed Swedish evolutionist, Professor Herbert Nilsson of Lund University stated that:

Mutations "are always...weaker" [Gene Depletion] and that in "free competition" [Nature] "they are eliminated" [by Natural Selection].

So while modern-day textbooks are falsely teaching students that: Mutations plus Natural Selection leads to Neo-Darwinian-Style Change, real science, which is based on millions of observations, reveals that the DNA Code Barrier plus Gene Depletion plus Natural Selection leads to no upward evolution of new kinds with new and beneficial genetic information having been added to their gene pool.

This is why I have been instructing people since 2002 that we can scientifically destroy Darwinian teachings in seven seconds flat. And that is if we take our time. Here again is how this is done:

"The DNA Code Barrier plus Gene Depletion plus Natural Selection makes Darwinian-style change scientifically impossible."

It is that quick and simple. In the business world it is what is called a *done deal*.

I learned how to refute Darwinism in seven seconds flat by reading about these three subjects and simply connecting the dots. Because this utterly destroys Darwinism, avid Humanists and their allies have attacked me continuously over my teaching. Their only other choice would be to admit that we were created—and this is untenable to them. Two facts jump out at me here: first, Darwinism is scientifically impossible; and second, name calling is the last bastion for those with no evidence to back up their position.

I received this note from a former Darwinian-trusting professor:

> "Russ, I was there last night and heard your presentation. I graduated from Case Western University in 1969 with a BA in Physical Anthropology. My senior thesis 'Anterior Tooth Reductions in Ramapithecus' was published in 1970 in the Journal *Primates*. I eventually got a PhD in History and was a Professor at the University of Arizona for seven years. I shared the lectern once at a national conference on eugenics with Stephen Jay Gould. In other words I've been there done that! In 1988, this know-it-all Jewish professor found Jesus. This is all to say that you did a marvelous job with your *50 Facts versus Darwinism in the Textbooks* presentation."
> Dr. Steven Yulish

Proverbs 9:8 Reprove not a scorner, lest he hate thee; rebuke a wise man, and he will love thee.

My seven-second rebuttal of Darwinism was confirmed by a genetic expert during late 2005. During November that year Dr. Jon Sanford, formerly of Cornell University, published a book titled *The Mystery of the Genome*. Dr. Sanford is a world-renowned genetic researcher with more than twenty patents on genetic research. In his writing he pointed out that genetic entropy (Gene Depletion) plus Natural Selection render Darwinism-style evolution to be scientifically impossible. This not only confirms what I have been teaching since 2002 but helps to take the heat off of me as one of the world's leading geneticists is now saying basically the same thing. I say *basically the same thing* because

Dr. Sanford did not add the DNA Code Barrier to his formula as I do.

I suspect that avowed Darwinists will claim that creatures began with the genetic data to evolve in place and have since lost the information due to genetic entropy. This is why I add the DNA Code Barrier which reveals that creatures never had the DNA to form different kinds of beings. They began with a wide range of variation within their particular kind of gene pool, but only the genetic information for variation within their own kind. They never had the DNA to form other kinds—DNA which they claim was subsequently lost. The DNA Code Barrier shuts the door on such false claims.

The debate over how the exquisitely-detailed genetic information found in the DNA molecule came to exist is a most important issue. This is also an issue that Darwinists will try to avoid. They have their hands full in attempting to explain how basic structures came about on their own. Failing miserably on this, trying to tackle where the complex specified information in the DNA came from, without God being involved, is way too much for them to handle. Although the secular crowd has no idea how the information could have come about on its own, they do agree that God could not have been involved. This is pure religious faith that flies in the face of real scientific facts.

As Richard Lewontin, former Harvard Professor of Zoology and Biology, explained:

> **"We take the side of (evolutionary) science... because we have a prior commitment to materialism. It is not that the methods...of science somehow compel us to accept a material explanation...on the contrary...for we cannot allow a Divine Foot in the door."** (From *Billions and billions of demons*; The New York Review; January 9, 1997; page 31.)

Neo-Darwinism is a religious belief that has been masquerading and undermining scientific research and scientific education for many decades.

A letter signed by dozens of scientists appeared in *New Scientist* on May 22, 2004 titled "Bucking the Big Bang." The letter included statements such as the following:

"The Big Bang theory can boast no predictions that have been validated by observation."

"Claimed successes consist of retrospectively making observations fit by adding adjustable parameters."

"The Big Bang relies on a growing number of never-observed entities, inflation, dark matter, dark energy...and can't survive without these fudge factors...In no other field of physics would this continual recourse to new hypothetical factors be accepted."

EIGHT

Darwinists Can't Overcome These Laws of Science

Naturalistic Darwinism is taught in America's public schools as if it were real science, but as we have pointed out, it is a refutable religious belief.

Here is even more valid evidence that will thoroughly refute the contentions long made by advocates of Darwinian-style (one kind of creature evolving into another kind of creature, such as apes evolving into humans) evolution.

In this and ensuing chapters I will discuss various subjects that Naturalistic Darwinists contend are proof of Darwinian-style evolution, and then counter their contentions with what real science reveals about that particular subject.

We will discuss a couple of biggies in this chapter: The First and Second Laws of Thermodynamics.

Where did the matter and energy come from?

The First Law of Thermodynamics states that matter or energy cannot be created or destroyed.

This is one of the most accepted and proven laws of science and is one of a number of serious problems for Naturalistic Darwinists who believe that *nothing* blew up in a Big Bang and everything eventually came into being over *billions of years* of time as a result.

Isaac Asimov referred to the First Law of Thermodynamics as "one of the most important generalizations in the history of science."

As simply as I can describe it, the First Law of Thermo-dynamics states that the total quantity of matter and energy in the universe is always constant, even though matter can be transformed into energy and energy can theoretically be transformed into matter.

This law, also referred to as the Law of Conservation of Mass and Energy, reveals that the universe could not have created itself as required by the various Big Bang cosmologies proposed by Naturalistic-oriented theories.

All the observed changes throughout human history, whether caused by natural forces or by mankind, have resulted from the rearranging of the energy and matter that had already been in existence. Never has an observed change occurred that was the result of the creation of new matter or new energy out of nothing.

The First Law of Thermodynamics is a major problem for all theories which attempt to explain how the universe developed itself without outside input. In other words, according to real science, the universe could not have burst into existence during a Big Bang from an infinitely tiny speck of matter.

A letter signed by dozens of scientists appeared in *New Scientist* on May 22, 2004 titled "Bucking the Big Bang." The letter included statements such as the following:

> **"The Big Bang theory can boast no predictions that have been validated by observation."**

> **"Claimed successes consist of retrospectively making observations fit by adding adjustable parameters."**

> **"The Big Bang relies on a growing number of never-observed entities, inflation, dark matter, dark energy...and can't survive without these fudge factors...In no other field of physics would this continual recourse to new hypothetical factors be accepted."**

Based upon examination of this issue from both sides of the fence – Naturalism and real science – it is clear that real science allows no room for a physically and mathematically impossible Big Bang event to make everything out of basically nothing.

So where did the matter and energy come from? Logically the cause of the existence of the universe's matter and energy had to have existed apart from the matter and energy found in the universe. The only viable explanation is found in the first five words of Genesis 1:1 which are that, "In the beginning God created…"

Throughout my school years

Talk of *The Big Bang* echoed throughout my school years, beginning at least by the third or fourth grade. It was just something else my young mind absorbed and accepted—and never challenged—until I realized the Bible is true word for word and cover to cover.

My co-author's son David has a blunt reply whenever someone asks, "Do you believe in *The Big Bang*?"

"Sure. God spoke and BANG the universe appeared."

However, this creation week event had nothing whatsoever to do with today's varying Big Bang theories. Still, David's statement pretty well captures the essence of the Creation report in the Book of Genesis, chapter one. If you unconditionally believe the Bible is the inspired Word of God – not just a portion of the Bible, but the entire book – you know Who created the universe, when He did so and how He did it.

Why do Darwinian-believing scientists call it the winding-up dilemma?

What else does real science reveal about the claim that a huge explosion – the so-called *Big Bang* – occurred sometime from 6 to 20 billion years ago, thus spawning the universe and billions of stars?

I will give you the shortest answer that I know and then explain what it means:

The galaxies are too tightly wound up!

Had such a big explosion actually occurred that many *billions of years* ago, stars should be evenly distributed throughout space. However, they are found in tightly-wound balls of stars or in spiral galaxies. They haven't had time to spread out.

According to secular, old-universe astronomy, the Milky Way is supposed to be more than ten billion years old. The problem for Darwinism is that this age is about twenty times longer than the spiral-shaped galaxies should have held to these tightly-wound patterns.

The stars that make up the Milky Way galaxy, the galaxy in which we live, rotate about the galaxy's center at various speeds. However, the inner stars rotate faster than the outer stars do.

The speeds at which these stars rotate are much too fast to support the suggestion that our universe is *billions and billions of years* old. Real science, to the contrary, which is based on the observation of the accessible evidence, shows that we live in a young universe rather than an old universe.

Here's why. The stars' rotation speeds, which have been scientifically observed, are so fast that if our galaxy were more than five hundred million years old it would have unwound and lost its spiral design. That is, the Milky Way galaxy would be a featureless disc of stars with great voids of space existing between them if it were as old as Naturalists claim.

Evidence that can be observed today by studying the Milky Way galaxy actually indicates a much younger universe. Much too young to provide the time required by the Darwinian beliefs.

In fact, the Hubble telescope has discovered millions of galaxies from as far as its telescope can see and all observed galaxies are either tightly-wound spiral shapes, as with the Milky Way, or in tightly-bunched balls of stars.

This is why Darwinian-believing scientists call it the winding-up dilemma.

Although scientists have known of this situation since the 1950's, the public does not hear much about it. Darwinian scientists refer to it as the "winding-up dilemma." They cannot explain it. And this dilemma is a problem that is not about to go away as it is a problem that applies to all other galaxies as well.

A recent example that bears this out was provided by the Hubble Space Telescope, which is mounted in an observatory that orbits about 353 miles above the earth. This powerful telescope found an extremely detailed spiral shape of stars inside of the central hub of what is known as the "Whirlpool" galaxy. This and many other such discoveries are in direct conflict to the predictions that are made by all *billions of years* old-universe cosmologies.

These scientific findings are a serious problem for these *billions of years* beliefs and for Darwinism, which has to have *billions of years* of time for Darwinian-style evolution to have any hope of validation.

Those persistent Secularists and their take on the Law of Entropy

The Law of Entropy is also known as the Second Law of Thermodynamics. This never-refuted law of observable science holds that all things in a natural setting lose usable energy and become disorganized over time. This scientific fact is a huge problem for Darwinists who teach that things evolve better and better over time, in direct conflict to this scientific principle.

Despite the overwhelming evidence that the Law of Entropy is in direct conflict with Naturalism, Secularists persist in claiming that their *open-system argument* overcomes this law, which is one of the most accepted of all scientific principles.

So how do Naturalists get around the Law of Entropy?

Well, to avoid this particular scientific roadblock to their belief that all life forms evolved from that first cell that spontaneously generated (violating the scientific Law of Biogenesis, which we will discuss in the next chapter), Secular Humanists employ their *open-system argument.*

Basically stated, this is the claim that an open system gets massive amounts of new energy added to the system so the Law of Entropy does not hold true in such a set-up. They apply the *open-system argument* to our solar system which receives tremendous amounts of raw energy from our sun. Humanists then claim that this addition of energy allows their belief to violate the Second Law.

However, there are many problems with this argument, and the fact is that the open-system argument does not help Darwinism. For example, if the Law of Entropy does not apply to our solar system, then how did we discover it?

Also, the much-observed fact is that raw energy tends to speed up both entropy and destruction. Take as a simple example a lawn chair. Let's say we bought a new lawn chair which came with a nice, thick-padded seat. After sitting on your patio for two years, the sunlight will have fairly well destroyed the seat cushion. The sunlight most certainly will not improve the cushion. Hoping that undirected raw energy can lead to better organization is like pouring gas on top of that lawn chair and lighting it on fire in the hope that the raw energy will improve its usefulness.

Dr. John Ross of Harvard stated:

> **"...the second law applies equally to open systems...the notion is that the law...fails for such systems. It is important to make sure that this error does not perpetuate itself."** (J Ross, *Chemical and Engineering News*, July 27, 1980, page 40.)

The important issue is the information in the system, not the energy available to the system. With regard to biology, it is a fact that raw, undirected energy like sunlight is incapable of producing the specified complex information found in living cells.

The Second Law of Thermodynamics, the Law of Entropy, is one of the most accepted laws in real science and is another major problem for Darwinism. The *open-system argument* does not overcome this principle of real science.

As you've seen in this chapter and will see in the following chapters, there are many challenges that Darwinists cannot overcome. As you read this report you will continue to grow in your understanding of why both Naturalism and Darwinism are in direct conflict with real science.

Former Harvard professor and Nobel Laureate George Wald stated why Darwinists continue on:

"I do not want to believe in God. Therefore, I chose to believe in that which I know is scientifically impossible: spontaneous generation [of life] arising to evolution."

NINE

The Law of Biogenesis

The Law of Biogenesis is another major roadblock to Darwinists. Real science does not wander off into what I call flights of fancy hoping to prove the unprovable. However, this is exactly what supporters of Darwinism do in trying to bolster their claim that life spontaneously generated itself from non-life.

And how does this claim hold up under the microscope of real science? Let's review the Law of Biogenesis.

One of the primary laws of science is the Law of Biogenesis. This law of real, observable science states that life can only come from living matter. In other words, non-living matter cannot produce living matter, an event which is required to get Darwinian evolution started.

To try and skirt around this scientific law, which dramatically and decisively refutes Naturalistic claims, Humanistic-biased textbooks teach that life spontaneously generated itself from non-living chemicals to become a "simple" single-celled creature, such as a bacteria cell.

Here's an example from a high school biology book published by Glenco in 1998. On page 324 kids read:

> **"All the many forms of life on Earth today are descended from a common ancestor, found in a primitive population of unicellular organisms."**

Either I'm becoming senile decades early or this statement is clearly being taught as if it were a verified fact. And what do they offer as observable proof? Well, amazingly, only three sentences later the text admits:

> **"No traces of those events remain..."**

Now, who is not thinking clearly?

It is a good time to repeat the second sentence in the opening paragraph of this chapter:

Real science does not wander off into what I call flights of fancy hoping to prove the unprovable.

To me, the statement "No traces of those events remain..." does not support Naturalistic or Darwinian philosophy as the text implies. Real science is very skeptical. The obvious question that comes to mind is, "If no traces of those events remain, how do we know those events ever took place?" The observable facts imply that there is no reason to believe that life spontaneously started on its own!

So why can't life spontaneously generate itself from non-life?

Real science provides us a good reason. Living systems are far too complex to have self-generated.

The field of biochemistry has discovered that bacteria cells are run by tiny molecular motors which operate, allowing the cell to perform its various functions. These microscopic rotary motors are called Bacterial Flagella and they are so complex and intricate that they can even change gears depending on how much weight they are either towing or pushing.

The Flagellum is made of about 40 different, highly-complex proteins and is known to be irreducibly complex. This term means that if any of the proteins were not entirely whole (with all of their specific left-handed amino acids and corresponding right-handed nucleotide sugars) and present in the exact order to first form this molecular motor at the exact moment for life to begin, life never could have spontaneously generated itself.

And to make matters even worse for Naturalistic Darwinism, the process of putting the proteins in the correct order to form the Bacterial Flagellum requires other molecular motors which are themselves irreducibly complex.

In simple, plain English there is no possibility that life began on its own.

It is of no wonder that the Law of Biogenesis has never been known to have been overcome, despite the very creative, non-scientific flights of fancy Darwinists wander off on.

Yet to the vast majority of earth's inhabitants this remains the biggest question of all:

Where did life come from?

Darwinists claim that life arose *billions of years* after a *BIG* unexplained explosion led to the formation of a big rock. Then oceans formed on the rock and from these primordial oceans life somehow arose, spontaneously generating on its own.

As a result of this naturalistic philosophy being imposed upon scientific researchers for the past fifty years, many roadblocks to real science have been erected. For example, real science reveals no known way that nature could allow life to develop by random chance from non-living matter. In fact, as already stated, the scientific Law of Biogenesis is simply that living matter can only come from already living matter. Yet all naturalistic origin-of-life theories require that this principle of real science be violated in order for life to have evolved on its own.

Due to this dilemma, most Darwinian-influenced biologists try to separate the origin of life, often referred to as chemical evolution, from discussions of biology. Their religious adherence to Naturalistic Darwinism leaves them no other choice but to ignore the scientific fact that life could not have begun on its own.

Yes, I used the word religious.

Why?

Because Naturalistic Darwinists have to take a leap of faith to believe their claims, which are in direct conflict with established scientific facts, principles, mathematical probabilities and all observations.

Modern-day science textbooks, written by Darwinian-worshipping authors, try to confuse students into thinking that life has been created from non-living matter by scientists in laboratories. For instance, experiments such as the one conducted by the Miller-Urey team in the 1950's have supposedly been able to produce some of the non-living chemical compounds found in living matter. However, despite the implied insinuation in the school books that life can spontaneously generate on its own, the fact is that humans have utterly failed to create life even in carefully designed experiments in the laboratory.

The Miller-Urey experiment came nowhere near creating life in the lab, much less revealing any way in which life could have overcome the Law of Biogenesis in a natural setting. That experiment and copy-cat tests since then have required massive amounts of intelligent input from the researchers. And this human input is built upon years and years of other scientists' findings. Then, in their closely-monitored laboratory, replicating conditions NOT even found in nature, they were able to produce a few of the 20 Non-Living amino acids that are found in living things. However, they had to immediately isolate the amino acids so that the conditions which made them would not quickly destroy them.

An even more significant problem for starting life in either a lab or in a natural setting is that of the 20 types of amino acids found in living matter, they can all form in either right- or left-handed versions.

Mathematically speaking, any such amino acids forming in such experiments or in a natural setting would end up with roughly a 50-50 mixture of both right- and left-handed varieties. Yet to form the proteins found in living systems, the mixture of amino acids must be left-handed only, with all right-handed nucleotide sugars (with only a few rare exceptions).

There is no known natural process that can produce only left-handed amino acids, much less with all right-handed nucleotide sugars.

The truth is that the Miller-Urey and other such experiments have not come anywhere close to producing life in the lab from non-living matter. The few amino acids they formed came in a mixture of right- and left-handed forms. So the actual fact is that they did not even come up with required non-living building blocks of life.

The fact remains that throughout the past fifty-five years, thousands of scientists, building upon years and years of previous scientists' research and experimental efforts, **have never succeeded in creating life from non-living matter.** The utter failure of scientists to create life from non-life in the laboratory is a major problem for the philosophy of Darwinian-style evolution.

What former Harvard professor and Nobel Laureate George Wald said about Darwinists continuing on is very revealing:

> **"I do not want to believe in God. Therefore, I chose to believe in that which I know is scientifically impossible: spontaneous generation [of life] arising to evolution."**

If scientists in a lab did get non-living matter to produce life would it prove that life could have begun on its own in nature?

Let us assume that one day, after spending billions of dollars and building upon years and years of research from thousands of other scientists, someone were able to get non-living matter to produce living matter.

Would this prove that life could have spontaneously generated without any intelligent being having been involved?

Hardly!

Real science reveals no way that life could have started on its own from non-living matter in a laboratory, much less in a natural setting. If scientists are ever able to create life from non-life in the lab, all it would prove is that it takes massive amounts of intelligent human input to create life.

So where did life come from?

This is the kind of question that can quickly multiply into a barrage of questions which Darwinists will keep trying to swat away by giving responses that beg even more questions.

Here is an example of what I mean:

Darwinism teaches that life somehow began on its own in the non-observable past. One well-known Darwinist was asked how life developed. He said, "From a crystal."

When asked, "Where did the crystal come from?" he stammered a bit and tried to say it must have come from some alien life form.

"And where did that alien life form come from?"

On and on the questions and responses went, never getting to a final answer.

Today, some evolutionary zealots, recognizing the impossibility that life could have spontaneously generated on its own, are suggesting that somehow life started elsewhere in the universe and was dropped off on earth by aliens, meteors or by some other unknown and unobserved process.

This simply shifts the Naturalistic problem to where it cannot be tested and to where it won't pose as big an embarrassment to secular scientists.

The point is that the facts do not support Naturalism because there had to be a starting point and an eternal starting agent.

Even though some major religions teach that EVERYTHING in the universe, including matter, is eternal, this would mean that matter would have to be self-maintaining. However, the Second Law of Thermodynamics displays entropy, the constant force which causes all matter and energy to become less and less organized over time, thus not maintaining itself (e.g., the sun WILL eventually burn out).

Conversely, the notion that NOTHING is eternal leaves us back at the unsolvable problem of the Darwinist who cannot logically answer the question of how the matter came into existence in the first place.

The Book of Genesis, the first book in the Bible, is supported by both logic and science. It very clearly reports how the universe was created, and by Whom it was created, that agent being our eternal Creator.

Yes, I know that the Bible is not a science book; however, God's Word does contain the true historical record of the universe. Among Scriptural accounts are the descriptions of many things which can be compared to scientific findings and none of these Biblical statements has ever been proven to be contrary to the observable facts – to real science.

This leaves the door open for the ultimate and obvious answer as to how the universe and life came into existence.

It is as stated in the Book of Genesis:

"In the beginning God created."

"Your theory is crazy, but it's not crazy enough to be true."

Niels Bohr, Danish scientist who was awarded the Nobel Prize for physics in 1922

TEN

Where Is That Missing Link?

Whenever I hear the words *missing link* I can't help but think of the time that I misplaced a cufflink and couldn't find it. Or the time at a bar-b-que when a link of prime sausage fell through the grill and into the fire.

Those were truly missing links!

So what about the *missing links* between ape and man?

Now here is an irony for you. As far back as the 1950's anthropologists using observable facts—real science—as their guide proved that Neanderthals were 100% human. Yet, even today, many Darwinists still claim them as *missing links*.

Is Neanderthal Man the missing link?

Neanderthal Man was the name given to human bones found in 1856 in Germany's Neander Valley. Many Darwinists still depict *Neanderthal Man* as a half-witted link between ape and man who lacked language skills and creative abilities.

Others say he was a dead-end in human evolution from our supposed ape-like ancestors.

However, the evidence has been around since the early 1900's which proves that Neanderthal was a variation of modern humans, and he has been officially reclassified as *Homo sapiens neanderthalensis*, a particular kind of modern man.

The renowned pathologist Rudolf Virchow presented evidence that the Neanderthal specimens which showed a hunched-over stance and other features were influenced by rickets, caused by a lack of vitamin D, and arthritis.

However, due to the overwhelming Darwinian bias, which has been undermining scientific education and scientific research for the past 100 years, his conclusions were held back from the public for half a century.

Another example of Darwinian bias was that it took longer than twenty years for the Field Museum of Natural History in Chicago to correct its display of Neanderthals even after their Neanderthal display was proven to be misleading Darwinian propaganda.

So much for honest *science*.

Darwinists still attempt to mislead anyone not on his or her toes.

For instance, although Neanderthal's brain size was slightly larger than modern man's, his brain is often claimed to have been of lower quality. However, this is just bias masquerading as science. Evidence reveals that *Neanderthal Man* lived at the same time as modern man and that they likely interbred with each other. The discovery of a Neanderthal hyoid bone, related to the voice box, which was no different than that of a modern human, has led many scientists to the conclusion that *Neanderthal Man* had speech abilities just like that of humans today.

Other evidence reveals that Neanderthals conducted religious rituals and were very creative. A Neanderthal toddler was unearthed in Syria with a flint tool resting at about the spot where the infant's heart had once beaten. Tools and jewelry such as pierced animal teeth and ivory rings were discovered with a Neanderthal fossil in a French cave in 1996. Well-designed and crafted stone tools and stone spearheads, as well as wooden spear shafts, have also been found. These finds and many others contradict the Darwinian claim that Neanderthals were a less-developed *missing link.*

It has been concluded that Neanderthals lived with other variations of modern humans in the Middle East and hybrids of Neanderthals and other humans are known from a number of areas around the world.

The only honest conclusion is that *Neanderthal Man* was 100% human.

What about Lucy – is she the missing link?

Lucy is another icon of the purported human evolution which Darwinists are reluctant to give up, even in the face of indisputable evidence that proves Lucy was nothing more than an ape conjured up to fool people into believing that Lucy was changing into a human.

Lucy is the name given to the much-promoted fossilized skeleton that was discovered during 1974 in Ethiopia by anthropologist Donald Johanson. Lucy has served as the poster child for Darwinism ever since.

It was claimed that Lucy walked upright, just like a human walks upright. Darwinists also claimed that they knew it was a *missing link* because the knee was "slightly bigger" than a normal ape's knee (proving that it was evolving into a human) and that its femur had to angle to the knee, just as a human femur angles to the knee.

However, what do the actual facts reveal?

Well, it is well documented that MANY humans have either slightly larger or slightly smaller than average-sized knees, and most tree-dwelling apes have angled femurs.

Also, according to another one of the world's best-known anthropologists, Richard Leakey, son of Louis Leakey, Lucy's skull was so incomplete that most of it is "imagination made of plaster of Paris." (*The Weekend Australian*, May 7-8, 1983, Magazine section, page 3.) Leakey stated in 1983 that no firm conclusion could be drawn as to what species Lucy belonged to.

And despite the fact that anatomists proved in 1987 that Lucy was just an ape, she is still claimed as the primary link between ape and man in textbooks around the world.

Scientifically speaking, Lucy is a member of a family of apes known as *australopithecines*, specifically *Australopithecus afarensis*. Pithecus means "ape," and as far back as 1987 honest scientists knew that Lucy was just an ape, and not a *missing link* between ape and man.

In 1987 Dr. Charles Oxnard, Professor of Anatomy and Human Biology at the University of Western Australia, wrote that although *australopithecines* were unique:

> **"Anatomists have concluded these creatures are not a link between ape and man, and did not walk upright in the human manner."** (*Fossils, Teeth and Sex — New Perspectives on Human Evolution,* Charles Oxnard, University of Washington Press, Seattle and London, 1987, page 227.)

Other skeleton finds of *Australopithecus afarensis* have shown that they had curved toes and fingers for gripping tree limbs.

The facts prove that none of the *australopithecines* are a transitional link between apes and humans. Lucy and the other *australopithecines* reveal nothing about supposed human evolution, yet still today, Lucy adorns high school and college textbooks, usually depicted walking upright, just like a human.

While failing to hold any proof for Darwinian change, Lucy does reveal the desperation of the Darwinian faithful, and how easy it is to be fooled by misleading textbook information.

When I think of the hocus pocus attempts of Darwinists to validate their belief system, my mind recalls those wandering salesmen of America's 1800's who traveled from town to town and settlement to settlement peddling everything from tin pans to special tonics guaranteed to cure "whatever ails you." They were sometimes not-too-fondly called "snake-oil" peddlers because of their bogus claims. I think this is also an apt description of Darwinists today. "Snake-oil" peddlers.

Modern textbooks have been claiming two new hominid discoveries (hominids are claimed to be the closest evolutionary link between ape and man). Kenyanthropus platyops, also known as Flat-faced Man, is one while the other is Sahelanthropus, also called Tomei Man.

Is Tomei Man Proof Of Darwinism?

The 2006 high school biology book titled *Biology* by K.R. Miller and J. Levine, and published by Prentice Hall, discusses these two Darwinian wonders on page 838. They state that Tomei Man "...is nearly seven million years old...older than any hominid previously known..."

Wow. That is quite an evolutionary find...or is it?

Tomei Man was first discovered in 2002. Later that same year *Science News* reported in its Oct. 2002 Volume 162 # 16, page 253, that:

> **"The specimen's teeth resemble..." those of the ape lineages and that 'it' "didn't walk on two legs..."**

In fact, *Nature* magazine reported in October 2002 that Tomei Man "...represents an ancient ape."

In other words, honest scientists knew when Tomei Man was discovered that it was just an ape.

A fair question is why are they teaching kids today that Tomei Man is proof for Darwinism?

The answer is that there is no viable proof to support Darwinian evolutionism! Thus they have to either admit that there exists no real evidence for this philosophical foundation for Secular Humanism, lending great support to the fact that we were created by God, or they must use known frauds to support their belief.

That brings us to "Flat-Faced Man."

All that was found of Flat-Faced Man was a small skull that was crushed into about 50 pieces. After "reconstructing" the pieces together it was announced that the face was "slightly" flatter than a normal ape's face, and this "slightly flatter" face was hailed as proof that it was becoming a human being.

Was this really all the "proof" needed to get Flat-Faced Man into public school and college textbooks? Yep.

What I have not seen printed in any of the schoolbooks is the interesting scientific fact that Flat-Faced Man would have stood two-feet tall!

That's right!

Our supposed closest relative on our ape-to-human tree was about knee-high to most adults (adult humans that is).

As you can see, it does not take much to qualify as evidence in support of Darwinian claims. Again, this is because there is no viable proof to support Darwinian evolutionism so they have to either admit that we were created, which they refuse to do, or keep grasping for straws to support their belief with whatever "snake oil" they can get people to buy into.

What is the mathematical probability that man evolved from an ape?

The more we learn about human DNA the more apparent it becomes that Darwinists are blowing hot air with their claims that some apelike creature lurks in man's ancestral closet.

Scientists who study such things say that human DNA molecules EACH contain enough hereditary data to fill a 500,000-page book. This data is translated by enzymes, all of which are encoded to protect against mutational defects. Both the genetic data that direct the formation of the enzymes and the enzymes that decode the genetic information had to be there from the very start.

For this reason alone, gradual evolution makes no sense at all.

Still, Darwinists teach that mankind evolved over long periods of time from some sort of an unknown apelike ancestor, often claiming that Human DNA is 98% the same as the DNA found in a chimpanzee.

Let's examine the truth and determine if this claim has any integrity or validity.

As far as the integrity of this claim, it was based on only a fraction of the total genome. The small portion of the DNA studied was that which is responsible for controlling the main body design. Since both people and apes have a torso, two arms and two legs, the fact that this portion of their DNA strand is rather similar was to be expected. To promote the claim that this proves we are 98% the same in our overall DNA as a chimpanzee is plainly dishonest and highly misleading.

So what about the viability of the claim that similar biochemistry proves that people evolved from apes?

Nature magazine has reported that as real science gets into the genome the wider the gap in genetic similarities becomes between man and ape. In fact, scientific studies reveal that there is at least a 7.7% difference between ape DNA and human DNA.

How big of a difference is that?

Well, consider that you contain about three billion base pairs of genetic information in every single one of your DNA-carrying cells. (Is it really logical that this kind of complex data formed without any intelligence behind it?)

Just a 7.7% difference between a human's genetic data and the gene pool found in a chimpanzee would mathematically require 231,000,000 beneficial and new genetic-information-adding mutations to take place in order to change a chimp into a human.

But never has even one such mutation been observed.

Keep these three facts in mind:

First, science knows of no way for nature to add appreciable amounts of new and beneficial genetic information to an existing gene pool.

Second, mutations are caused by the sorting or the loss of the parents' genetic data, not by the creation of new and beneficial DNA.

Third, so many mutations are fatal that there is no mathematical possibility of stringing together 231 million in a row without exterminating the potentially evolving individual.

So why don't Darwinists stop promoting misleading information and just bring out the real links, the non-missing links that prove we evolved from a single-celled creature to an ape to a human?

Because there is no indisputable evidence to back up the secular-biased claims. In fact, there are not even any semi-viable examples of a link between a single-celled paramecium to a parrot, to pine trees, to pickles or even to Darwinian-proselytizing professors.

Zero. Nada. None.

The "links" are all missing and they are all missing because they never existed.

"If my theory be true, numberless intermediate varieties...must assuredly have existed."

Charles Darwin

Dr. Duane Gish hit the proverbial nail on the head when he defined Darwinism as:

"The sustenance of fossils hoped for, the evidence of links unseen."

ELEVEN

And Where Are Those Intermediate Fossils?

"Where's the beef?" is a famous question in advertising history. Three old ladies were eyeing some big, fluffy hamburger buns when one of them suddenly, gruffly demanded what all three of them wanted to know, which was why the hamburger patty itself was so small. So she asked, "Where's the beef?"

According to real science, a version of that same question is appropriate to challenge advocates of Darwinism with:

Where are the intermediate fossils?

As Darwin admitted, numberless intermediate varieties must have existed if everything evolved from a single-celled creature. Unfortunately for Darwinists the query above is a question they cannot honestly answer.

Millions of fossilized fish and fossilized amphibians have been found, yet never has an intermediate fossil of a finned fish evolving into a legged amphibian been located. Although claims for transitional candidates are often made, never do the claimed evolutionary links stand up to true scientific scrutiny.

I repeat:

Many frauds have been offered up and many dubious candidates for transitional kinds have been presented only to be refuted at a later time, however not a single transitional fossil (one kind of creature, such as a fish, evolving into another, such as an amphibian) has ever been verified among any of the millions of fossils that have been found.

Charles Darwin realized that the fossil record, as it was known during his lifetime, did not support the predictions made by his own theory of evolution. However, he hoped that in the future researchers would begin to find the overwhelming number of *missing links* that should have existed in the fossil record if his thoughts were accurate.

Darwin wrote:

> **"Why is not every geological formation and every stratum full of such intermediate links? Geology assuredly does not reveal any such finely graduated organic chain; and this is the most obvious and serious objection which can be urged against the theory."** (C. Darwin, *Origin of Species*, 6th edition, 1872; London, John Murray, 1902, page 413.)

As one example of the much-promoted *missing links*, the April 2006 issue of the scientific journal *Nature* reported the discovery of several well-preserved specimens of fish in sedimentary layers of siltstone in Canada.

These fossilized fish specimens are called *Tiktaalik roseae*, and like other lobe-finned fish, were declared to be from the late Devonian age. Accordingly, Naturalists who deny the global flood believe these fish to be somewhere between 359 and 385 million years old. The discoverers further claimed that these "represent an intermediate between fish with fins and tetrapods with limbs."

By claiming the fossils to be the long sought *missing link* the *finders* were immediately vaulted to world-class status amongst the Darwinian-worshipping ranks of scientists who are in desperate need of evidence in support of their failing religious belief.

However, the fact is that *Tiktaalik roseae* was 100% fish.

In a review article on *Tiktaalik roseae*, which appeared in the same issue of *Nature*, fish evolution experts admitted that, in many aspects, *Tiktaalik roseae* "are straightforward fishes: they have small pelvic fins, retain fin rays in their paired appendages and have well-developed gill arches, suggesting that both animals remained mostly aquatic."

In other words, the hailed *missing link, Tiktaalik roseae*, was just a fish.

However, due to their evolutionary leaning, the writers did weakly try to claim that *Tiktaalik's* "skull has a longer snout," feebly supporting the notion that perhaps "a longer snout suggests a shift from sucking towards snapping up prey..."

The renowned Darwinian Marxist Stephen Jay Gould wrote:

> **"The absence of fossil evidence for intermediary stages between major transitions in organic design, indeed our inability, even in our imagination, to construct functional intermediates in many cases, has been a persistent and nagging problem for gradualistic accounts of evolution."** (S. J. Gould, *Evolution Now: A Century After Darwin*, ed. John Maynard Smith, New York: Macmillan Publishing Co., 1982.)

Do any fossils prove that Darwinian-style evolution ever occurred?

For what reason I do not understand, Darwinists have been desperately attempting to convince people that dinosaurs evolved into birds. So what evidence supports that birds evolved from dinosaurs?

Never has an intermediate fossil of a reptile evolving into a bird been found. Thousands of fossilized reptiles and birds are documented, yet not a single transitional fossil has been discovered that supports, much less proves, that one kind morphed into the other creature.

The amount of transitional forms involved in the supposed evolutionary transformation from reptiles to birds, over *millions of years*, should add up to an incredible number of links. Although the only fossil evidence presented has not held up to even scant scientific scrutiny, it still has misled billions of people into believing in Darwinian-style evolution.

After one embarrassing candidate that was passed off in American museums and media as a reptile-to-bird missing link was proven to be a fraud put together by a cash-poor farmer in China, "The 'Missing Link' That Wasn't" by Tim Friend appeared in *USA TODAY* on February 3, 2000.

In the article, Friend wrote:

> **"...this 'true missing link...between dinosaurs and birds'...sprouted its...tail not 120 million years ago but only shortly before being smuggled out of China...has children believing in feathered dinosaurs that never existed, prominent scientists calling each other names and two respected science publications under assault..."**

Unfortunately, these types of fraudulent claims hit the newsstands about once a year. After being used to mislead millions of people they then quietly disappear.

The primary candidate presented by Darwinists as the missing link between reptile and bird is *Archaeopteryx*. There is no doubt that *Archaeopteryx* was an odd bird but there is also no honest doubt that it was 100% bird. It had well-developed feathers for flight, and feathers are very complex structures.

Researchers now know for a fact that reptile DNA does not contain the information to form feathers. The dagger through the heart of Darwinism here is that real science knows of no way for nature to add appreciable amounts of new and beneficial genetic information to an existing gene pool, much less the millions of pieces of data required to change a reptile into a flying bird.

Another crushing blow to the claim that *Archaeopteryx* was a missing link between reptiles and birds is that scientists have found several fossils of completely modern birds below the strata layers which contained *Archaeopteryx*. "...scientists found bird bones ...farther down the geologic column than *Archaeopteryx*..." (*Nature* 322; 8-21-1986; *Science* 253; 7-5-1991.)

Since Darwinists believe those layers of rock formed slowly and that the fossils found in those strata represent our evolutionary paths, finding modern birds in layers below those that contain *Archaeopteryx* should have removed him from consideration as any sort of transitional kind.

The overall fossil record is an enemy to Darwinism.

All higher kinds of plants and animals appear abruptly in the fossil record with no transitional types linking one group to another.

The late paleontologist, Dr. Colin Patterson of the British Museum of Natural History, wrote a book titled *Evolution*. In reply to why he had not included any pictures of transitional forms in his publication, Dr. Patterson wrote:

> **"I fully agree with your comments about the lack of direct illustration of evolutionary transitions in my book. If I knew of any, fossil or living, I would certainly have included them...I will lay it on the line, there is not one such fossil for which one could make a watertight argument."** (C. Patterson, from a letter to Luther Sunderland, April 10, 1979, as published in *Darwin's Enigma*, Green Forest, AR: Master Books, 4th ed. 1988, page 89.)

To his credit, Charles Darwin was a blunt realist. He fully realized that if his theory of evolution were true there should have been abundant proof to back him up. Darwin stated on page 211 of his first book, *On the Origin of Species by Means of Natural Selection, or the Preservation of Favoured Races in the Struggle for Life*, that:

> **"If my theory be true, numberless intermediate varieties...must assuredly have existed."**

I certainly agree with Darwin's statement. If everything in the world had evolved from that supposed first living cell, then the fossil evidence would be overwhelming. There should be nothing to debate. But the application of real science, which means stringent observation and examination of the visible evidence, has uncovered no such proof.

The complete lack of transitional fossils refutes Darwinian evolution by itself. Had a living cell overcome the Law of Biogenesis and spontaneously generated itself from non-living matter, and had everything evolved from that supposed single cell, the fossil record would contain millions—even billions—of missing links. In fact, we would see numberless transitional kinds flopping around the earth or wiggling in the seas today. But the links are missing, refuting Darwinism by their absence.

Prior to trying to comply with Darwinism, science had been based on knowledge derived from the study of the evidence.

Back in the 1930's Richard Goldschmidt had the same problem that Darwinians have today. He had no evidence that Darwinian change had ever occurred. Goldschmidt developed the *Hopeful Monster* theory to try and cover the gaping holes in the fossil record where all things are found as completed kinds, not intermediate or transitional types of plants or animals as Darwin had expected would be found, if his theory were correct.

The Hopeful Monster theory basically claimed that there was no evidence of Darwinian change in the fossil record because Darwinian change occurred virtually instantaneously. Goldschmidt even suggested that perhaps reptiles laid eggs and birds popped out, leaving no evidence behind! It is okay for you to laugh about the Hopeful Monster theory; almost everyone else did.

Well, fifty years later there was still no evidence of Darwinism having occurred so Niles Eldridge and Stephen J. Gould changed the Hopeful Monster theory just slightly, but gave it a much more scientific-sounding name.

Now the theory which explains why there is no evidence of Darwinian change in the fossil record is referred to as *Punctuated Equilibrium.* Although this sounds rather daunting, what it basically means is that there is no evidence of Darwinian change in the fossil record because Darwinian-style evolution occurred in such short geological bursts of time that no evidence was captured in the fossil record—in other words, a reptile laid an egg and a bird popped out, but it was so fast we missed it!

But isn't science supposed to be knowledge that has been derived from the study of existing evidence?

So what became of science being based on knowledge derived from the study of the facts? Well, the total lack of viable evidence in support of Darwinism required either: a] an abandonment of Darwin's theory (which would have put a torpedo through the side of Secular Humanism, while leaving Biblical Creation as the only viable alternative); or b] a complete change in the definition of the word *science*.

As unimaginable as it seems, the word *science* was redefined to fit the religious philosophy of Naturalism. Today the word *science* is defined as the study of "naturally occurring phenomenon." The problems with this redefinition of the word are many, including undermining scientific study, scientific education and the faith of billions of people in the Lord Jesus Christ.

The fossil record has also been a total embarrassment to Secular Humanists whose beliefs are today founded on Darwinian-style change being true, removing God from the picture. Still, the refutable beliefs of secular atheists have been taught in America's public schools since 1963 as if they were scientific facts. This has led to the removal of God from American society and to the loss of America's heritage and the continued erosion of its citizens' freedoms as we reveal in our first report, **The Theft of America's Heritage: Biblical foundations under siege: a nation's freedoms vanishing.**

The United States of America is now reaping the fruit from these false teachings throughout its society. The sexual revolution (1966-1967) leading to the radical women's liberation movement, the homosexual movement, legalized abortion (1973), the tremendous increase in sexually-transmitted diseases, the explosion of pornography and more, including the destruction of the basic family unit, are all direct fruits which quickly followed the replacement of the teaching of Biblical Creation with the false science of Darwinian evolution in America's public schools in 1963.

Again, prior to trying to comply with Darwinism, science had been based on knowledge derived from the study of the evidence. Darwinism is not science; it is a religious belief which has been employed to promote Naturalism and Secular Humanism (atheism) while greatly undermining scientific research, scientific education and the faith of billions of people.

Still, the fossil evidence is missing because Darwinism never took place. It is a refutable religious belief.

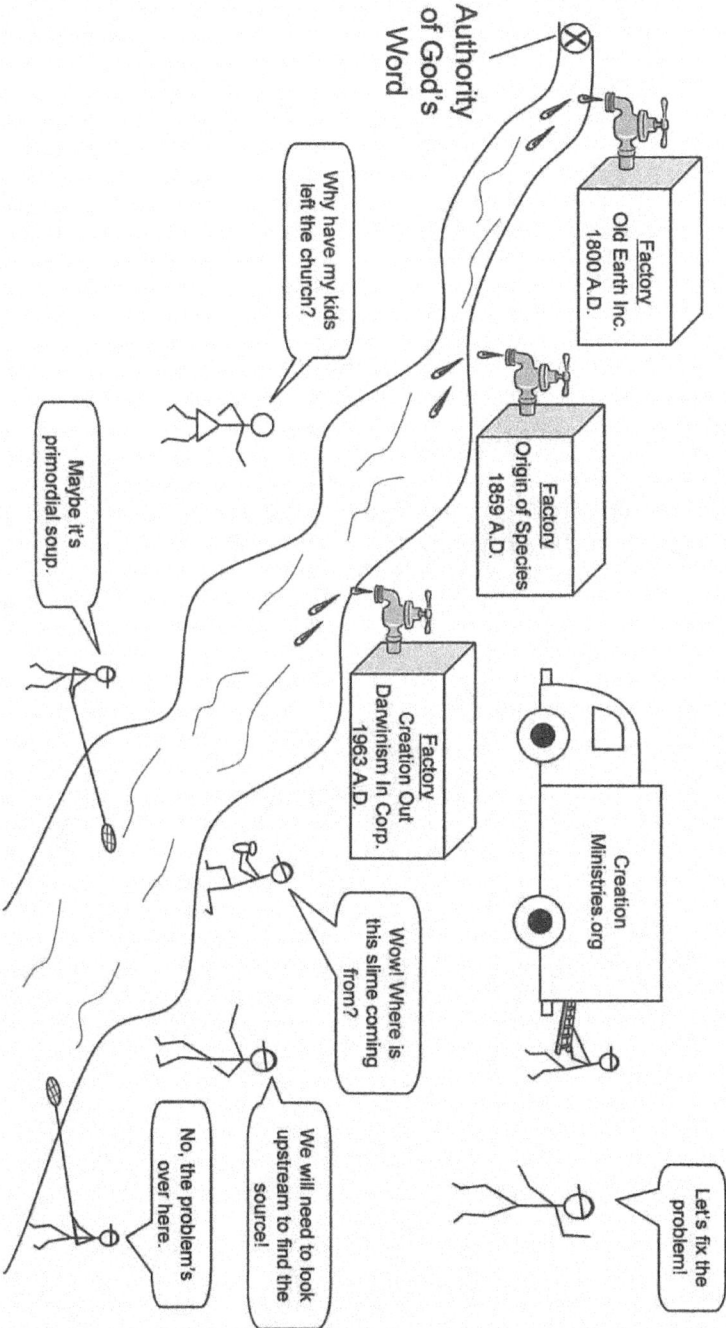

"Darwinists are experts at making drawings of things that never existed to support their theory that never took place."

Often repeated by Russ Miller
Original source unknown

TWELVE

Masters of Artistic Fraud

As a kid in school I was mesmerized by the drawings of incredible creatures. Some of those drawings even showed the purported steps of evolution that various animals went through to become like they are today.

I didn't question those drawings. After all, they were in the school textbooks so they must be true, right? My young mind was a sponge and it accepted what was poured into it. Looking back it is easy for me to see how I became a Theistic Evolutionist, believing in a God who would have needed *millions and billions of years* to bring all His creatures and those quadrillions of stars and everything else into existence.

It wasn't until I was presented with the facts which exposed the Darwinian frauds and the evidences in support of Biblical Creation and the global flood when I was on the threshold of middle age that I realized that the Biblical account of creation actually could be true. It was at that point that I came to the realization that Darwinists are experts at making drawings of things that never existed to support their theory that never took place.

For instance, Darwinists show drawings of a human in the embryonic stage and claim that humans have gill slits, or gill pouches, from our past evolutionary stages. However, these are not gill slits or gill pouches. These are simply folds in the skin which eventually develop into the organs in our throat and neck area.

This is a takeoff of Ernst Haeckel's fraudulent work from the 1860's. Haeckel had read Darwin's book *On the Origin of Species by Means of Natural Selection or the Preservation of Favored Races in the Struggle for Life* shortly after the book was published in 1859.

Same problem then as today.

Haeckel had the same problem that Darwinists are confronted with today; he could not find any real evidence to support the Darwinian belief. So Haeckel improvised and became one of the Darwinian pioneers for inventing evidence that never actually existed except in their wishful thinking and fraudulent examples.

What a mouthful!

Haeckel came up with the "Biogenetic Law," also known as the "Theory of Recapitulation," that "ontogeny recapitulates phylogeny."
I know. It sounds like a mouthful.   In reality, it's a classic example of pure hogwash.
Simply stated, this is the teaching that you go through your past evolutionary stages while in your mother's womb.
What Haeckel did was to take a human in the embryonic stage and draw copies of the human with slight changes in the drawings. Haeckel then labeled his drawings as human, salamander, chicken, fish, and other various critters all looking almost identical.   Then he announced his "finding" as proof that we go through our past Darwinian-style changes while in our mother's womb.
Embryologist Dr. Michael Richardson stated:

> **"What he (Haeckel) did was to take a human embryo and copy it...these are fakes."** (*The Times,* London, August 11, 1997; page 14.)

Please remember that Darwinists are experts at making drawings of things that never existed in order to support their theory that never took place. Keep this old saying in mind whenever you encounter the supposed Darwinian "proofs" and you will start to realize that their "evidences" are almost always drawings of how they think the evidence would appear if any actual evidence were ever found. Thumb through your local high school's biology book and you will find examples that bear out the truth I am telling you.

Haeckel was proven to be a fraud in the 1870's yet variations of his 100% false "theory" are still taught in colleges around the world today. The simple fact is that fraud in the nineteenth century is still fraud in the twenty-first century and having to draw things that never existed only supports that Darwin's theory never took place.

Consider the Ambulocetus drawings.

Darwinists claim that Ambulocetus is the missing link between land mammals and whales. They also claim that the whale evolution series is solid proof of Darwinism in action.

Sadly, this *is* Darwinism in action, but not because it is based on actual scientific evidence of Darwinian-style evolution having taken place. Unfortunately, as I said and will say again, Darwinists are skilled at drawing things that never existed to promote their theory that never took place, and the whale evolution series is a fine example of such "Darwinism in action."

Drawings depicting an extinct land mammal are posed next to Ambulocetus which is followed by a whale. This is a deliberate attempt at creating and leaving a distinct impression in the minds of the book's readers that Darwinian-style evolution is a scientific fact.

However, the much-promoted half-land-mammal, half-whale Ambulocetus, beautifully drawn in full-color depictions, never actually existed. The mythical creature put forth as the long-awaited missing link of one kind of an animal evolving from a different kind of animal was in reality made up of bones that were found in separate strata layers and in different locations!

That is correct: the bones aren't
even from the same animal.

In fact, whatever animals the bones are from did not even live at the same time, according to the evolutionary interpretation of the strata layers! And even with these deliberate shenanigans, only about 24% of a skeleton was made up and it contained no pelvic girdle.

In other words, no one could know whether the creature would have swam, walked or crawled, yet drawings in modern-day schoolbooks show it as the missing link between land mammals and whales. This fraudulent "proof" of Darwinian change has led millions of people into believing that there is no Creator and not any need to allow for God in their lives.

Darwinists truly are skilled at drawing things that never existed to promote their theory that never took place; and Ambulocetus, the non-existent missing link between land animals and whales, is nothing more than a ploy to mislead many people.

Do any textbook drawings or bone arrangements prove Darwinism?

I clearly remember those drawings of horse bones in textbooks when I was in grade school, high school and college. Sadly, this is another example of Darwinism in action, but not because it is based on actual scientific evidence of Darwinian-style evolution having taken place.

As I've hammered in this chapter, Darwinists are skilled at drawing things that never existed to promote their theory that never took place, and the horse evolution series is a classic example of such "Darwinism in action."

Textbook drawings and actual bone arrangements in museums depict a supposedly extinct "ancient" horse next to a slightly larger horse next to the modern-day horse. This is a deliberate attempt at creating and leaving a distinct impression in the minds of the book's readers and museum visitors that this is solid proof of Darwinian-style change.

However, the much-promoted horse-fossil evolution series, found beautifully drawn in full-color depictions in various textbooks, or magnificently arranged with actual bones going from the small "ancient horse" to the large modern horse has never been found in the order that they are presented.

In fact, the supposed modern-day horse has been found in strata layers below the rock layers that contain the supposed ancient horse. Well, according to the evolutionary interpretations of the strata layers, that puts the modern-day horse on earth prior to the supposed ancient horse that the modern-day horse was supposed to have evolved from!

Furthermore, the order in which the horses are arranged has never been found in strata layers anywhere in the world. Nowhere have the fossilized horse bones matched the supposed evolutionary path that Darwinists are teaching in the schools and museums.

Nowhere. Never.

Incredibly, those same, much-promoted drawings and museum displays depicting the horse evolution series are still being presented to young, impressionable students as scientifically valid documentation of Darwinian-evolution.

Another sad but interesting note is that of all the museum displays that I am aware of which depict the horse evolution series showing the small horse next to the larger equine next to the larger modern horse are all made from the bones of modern horses!

That is correct.

They have taken the bones from a one-month-old horse, the bones from a three-month old horse, a year-old horse and an adult horse and lined them up in order by size, and then employed them to fool millions of people into believing that Naturalists actually have proof of Darwinism in action.

Again, this sadly is Darwinism in action, but not because it is based on actual scientific evidence of Darwinian-style evolution having taken place. The bottom line remains unchanged – which, again, is that Darwinists are skilled at making up or drawing things that never existed to promote their theory that never took place.

And the horse evolution series, which is a perfect example of this deliberate deceit, has been employed to mislead many people.

Nebraska Man was an example of incredible imagination.

Nebraska Man was another work of Darwinian creativity. All that was found of Nebraska Man in the early 1920's was a piece of a broken tooth. And from that broken tooth evolutionary artists reconstructed Nebraska Man, his family and even the wood and stone tools that he supposedly worked with. From a piece of a broken tooth!

It was later proven that the tooth came from an extinct peccary, an animal somewhat resembling a pig.

I could give a lot more examples of Darwinists' artistic frauds, but I think you get my point:

> **Darwinists are experts at making drawings of things that never existed to support their theory that never took place.**

"I reject evolution because I deem it obsolete; because the knowledge, hard won since 1830, of anatomy, histology, cytology, and embryology, cannot be made to accord with its basic idea. The foundationless, fantastic edifice of the evolution doctrine would long ago have met with its long-deserved fate were it not that the love of fairy tales is so deep-rooted in the hearts of man."

Dr. Albert Fleischmann, University of Erlangen in Germany

THIRTEEN

Homeo Box Genes, Bacteria and Viruses

What do Homeo Box genes, bacteria and viruses have in common in the creative minds of Darwinists?

If you thought *mutations* you get an A.

Let's discuss Homeo Box genes first.

Darwinian-biased scientists claim that mutations in Homeo Box genes lead to the evolution of new and improved kinds of creatures.   However, the observable facts reveal this to be an erroneous conclusion.

Homeo Box genes are known as Hox genes for easy reference. Hox genes are control genes. Just as a traffic cop directs the flow of automobiles to where they need to go, Hox genes direct genetic information to where that particular data is required for the development of a specific portion of a given organism.

Take a frog as an easy-to-imagine example. The frog has four legs and in the frog's gene pool is found the genetic information to form its left front leg, its right front leg, its left rear leg and the frog's right rear appendage. This genetic blueprint is extremely detailed and will produce the appropriate limb, which will allow the frog to live its life productively, for a frog.

However, how does this coded hereditary information know where to form the appropriate leg?

This is where the Homeo Box gene comes into play.

The Hox gene is the traffic cop for the various genetic information in a given plant or animal. The Hox gene directs the genetic data to form the frog's left rear leg in the proper place on the frog's body. Without the Hox gene, the frog's genetic data would not know where to properly form the limb.

A mutation in the Homeo Box gene, which is the result of the mix-up or loss of its pre-existing genetic information (data inherited from its parents), can cause all sorts of problems and lead to monstrosities which usually soon die. If not immediately fatal, most Hox gene mutations lead to plants or animals which are the weakest of their specific kind in nature, and to their early removal via Natural Selection.

Think back to our discussion in Chapter Seven:
Refuting Darwinism in Seven Seconds Flat!

> **Natural Selection is the scientifically observable process by which weaker kinds are removed by the competition for resources in a natural environment.**

A Hox gene mutation in the amphibian we are using for this example can result in the frog's genetic information being steered in the wrong direction, thus forming its left rear leg in the wrong location.

If you have ever seen a picture of a fruit fly, often used in mutation experiments in laboratories, with one of its legs protruding from its forehead, this was most likely the result of a mutation in one of its Homeo Box genes. A similar mutation in our frog's Hox genes may result in the frog's left rear leg coming out from the middle of the poor amphibian's back, and there is nothing beneficial about this kind of mutation.

The leg which has developed in the wrong location will not be functional as it will not have the skeletal muscles or nerve connections necessary to be useful. The improperly placed leg will hang useless off of the poor frog's back, yet burning up the frog's energy and resources to maintain it. This will make the frog the weakest of the frogs and in a natural setting, our frog will be the one most likely removed by Natural Selection.

What do Darwinists claim about bacteria?

Darwinists claim that bacteria becoming resistant to antibiotics such as penicillin is proof that they are evolving into something that is becoming better and better in Darwinian fashion.

In fact, the *Teachers Guidebook* which was put out by the National Academy of Sciences (NAS) in 1998 claimed, in a section titled *Teaching About Evolution* on pages 16-17, that continual evolution produces bacteria that are resistant to antibiotics.

However, observable scientific facts reveal that this has nothing to do with the Darwinian-style evolution of new kinds of organisms with appreciable amounts of new and beneficial genetic information having been created and added to their parents' pre-existing gene pools.

It is a scientific fact that bacteria can become resistant to antibiotics via a mutation that causes an anti-penicillin enzyme to mass produce.

It is also a fact that Darwinists claim that the mutation that causes bacteria to be immune to penicillin is proof for their evolutionary belief.

And what does real science reveal?

Real science reveals that this mutation, like all observed mutations, is caused by the sorting or loss of the parent bacteria's pre-existing DNA, and not by the creation of new and beneficial genetic information.

Let us be completely fair. Although this genetic loss has nothing to do with Darwinian change, does it actually convey a benefit to the bacteria?

Well, if that mutated bacteria were in a person, and if that person were in a hospital, and if that hospital's staff were administering penicillin to that patient, then in that one particular situation the mutation would provide that one bacteria with a benefit.

However, everywhere else in the world that poor little bacteria would be using up much of its energy and resources manufacturing the unneeded anti-penicillin enzyme. This would make it the weakest bacteria around and lead to its removal via Natural Selection.

Bacteria becoming resistant to a particular antibiotic has nothing to do with Darwinian-style evolution. That is, there have been no new kinds of organisms with appreciable amounts of new and beneficial genetic information created and added to their parents' pre-existing gene pools.

I repeat: these weakened bacteria have nothing to do with Darwinian change.

What do they claim about viruses?

Darwinists make the same claim about viruses as they do about bacteria. And in both cases these claims are not supported by scientific evidence. In the decade I've devoted to studying scientific research as it relates or does not relate to Darwinism, I have been overwhelmed by how non-existent the evidence is which supports the Darwinian claims.

I have found no undisputable evidence in support of Darwinian change.

None.

You might say that the evidence supporting Darwinism is "underwhelming."

If Darwinism is science which should be accepted and not questioned, as Secular Humanists claim, then there should be unlimited scientific proof that those claims are valid. This vacuum of evidence speaks volumes.

Again, Darwinists claim that the AIDS virus or the Bird-Flu virus may evolve, through mutations, into something that is better and stronger until they become immune to current medical treatments, leading to a worldwide health pandemic.

While it is true that one of these viruses could suffer an alteration that could cause it to be uncontrollable by present treatments, observable scientific facts reveal that this alteration in the virus has nothing to do with Darwinian-style evolution. That is, there is no observable evidence that new kinds of organisms with new and beneficial genetic information have been created.

It is a scientific fact that mutations do occur.

It is also a scientific fact that these mutations are caused by Gene Depletion, just as we have pointed out. As I previously said, Gene Depletion (also referred to as Genetic Entropy) is the observable principle that mutations are caused by the sorting or loss of the parents' beginning genetic information, and not by the gain of new and beneficial genetic data.

Both bacteria and viruses multiply rapidly, accumulating genetic losses due to mutations. Realize that these are simply micro-changes within the same kind of virus or bacteria, not the evolution of some new and distinct organism which was caused by the addition of massive amounts of new and beneficial genetic data as Darwinists would have you believe.

The fact is that anti-bacterial or anti-viral treatments are developed to recognize and attach to a particular protein on the pathogen. It does this by recognizing the protein's shape. If a mutation (a micro-adaptational change) caused by the recombination or the loss of the starting genetic information alters the shape of the particular protein that the antibiotic recognizes and attacks, it can make the vaccine useless.

This micro-change, caused by the sorting or the loss of the pre-existing genetic information, can alter a small health risk into a worldwide pandemic. Nevertheless, such changes in either bacteria or viruses have nothing to do with the Darwinian-style evolution of new kinds of organisms with new and beneficial genetic information having been created and added to their parents' pre-existing gene pools which eventually creates a new kind of being.

Insects or other pests such as rats becoming unaffected by poisons are other good examples of such micro-changes caused by the mutational sorting or loss of genetic data. An alteration caused by such a loss can cause vermin, such as a cockroach, to become immune to a particular insecticide. While the poison kills off the non-mutated cockroaches, the mutant survives the toxin and passes on the genetic loss to its offspring who are also immune to that particular insecticide. A beneficial change indeed but not resulting from the gain of genetic information and not leading to Darwinian change. This is just another micro-adaptation, one kind bringing forth after its own kind, just as the Word of God tells us will occur.

As of this writing, real science knows of no viable example of nature adding appreciable amounts of new and beneficial genetic data to an existing gene pool. The simple fact is that if the millions of various kinds of species found alive on earth today had evolved according to the Darwinian model, from some single-celled being, Darwinists should be able to easily present millions of examples of how nature can add appreciable amounts of new and beneficial DNA to create the changes.

Yet they cannot show a single viable example, which is ironic because Darwinian-worshipping Secular Humanists control the textbooks used in America's public schools, as has been the case for several decades, and their textbooks present Neo-Darwinism as scientific fact that should be accepted and not questioned.

However, real science does not support Humanists' claims. The evidence in support of the Darwinian tale is truly "underwhelming."

Proverbs 15:13a
A merry heart maketh a cheerful countenance.

_____

Another way to put it is "laughter is good medicine." Our hope is that the chapter you're about to read will crease your face with smiles.

Jim Dobkins, co-author of this report

_____

"In fact, if similarities support Darwinism the textbooks should be teaching that we evolved from sunflowers because our cytochrome C is closest to that of a sunflower. In fact our eyes are closest to that of the octopi while human skin is closest to that of swine. I really think that if Darwinists are going to teach kids that similarities prove that we evolved from a rock that they should note that we were once root nodules, since human hemoglobin structure is closest to that of root nodules. Human milk is closest to that from donkeys while...well you get the picture.

The obvious fact of the matter is that similarities in biochemistry have nothing whatsoever to do with us having evolved from other beings. In fact, similar biochemistry is crucial for our survival. If we did not have similar biochemistry with other plants and animals we would not be able to digest them. In other words, without such commonalities we would only be able to eat other people. Similar biochemistry is proof in favor of our Intelligent Biblical Designer as opposed to evidence favoring Darwinian change."

Russ Miller

FOURTEEN

Why Not Snow Cones?

I have a confession to make. It's for your eyes only, so don't you dare tell anybody – at least not until you have the confidence to openly and professionally discuss what you are about to learn.

All this discussion about missing links that do not exist, intermediate fossils that do not exist, and drawings of critters that never existed has convinced me of something potentially momentous.

Are you ready for this?

Are you sitting down, braced for this revelation?

I hate to admit it but I am on the verge of becoming the greatest Darwinist of all time! I will achieve this dubious distinction by being, to my knowledge, the first person to actually publish the following hypothesis which contains actual proof of Darwinian-style change having occurred.

Hear me out and see what you think.

Watermelons and clouds

It has long been known that watermelons are about 96% water while clouds are about 100% water. Obviously there is a close biochemical makeup which must certainly prove that they evolved from a common ancestor. However, that 4% difference also strongly indicates that there are some missing links involved, so I and many others set out to find them.

Through years of investigation, we have found that jellyfish are 98% water while snow cones are 99% water and herein lies the first evolutionary tree of life which is actually backed by observable facts.

In this hypothesis it is suggested that a bacterium overcame the scientific Law of Biogenesis, as well as all mathematical probability, and mutated its way, overcoming the scientific principle of Gene Depletion, to become the lowly watermelon. Next, over *millions of years* of time, the watermelon branched off into separate evolutionary pathways which included both jellyfish and snow cones.

I believe that the jellyfish have been an evolutionary dead-end (although there are a few researchers who require further study) and remain as various jellyfish today. However, evidence from the scientific field of hydrology proves that ice melts to water and further warming will turn the water to vapor which rises during the process of evaporation.

This was what clicked on the light bulb.

I, as well as many others, believe that the snow cone melted and then evaporated, rising high into the sky to form the first cloud. And here is the best proof for Darwinian-style change having taken place.

By drawing an evolutionary tree of life, with nice colorful lines connecting the watermelon to the jellyfish in one pathway, and other brightly colored lines going from the watermelon to the snow cone and on to the cloud in another evolutionary route, I believe we have the first actual proof of Darwinian change having occurred!

You should be thinking, "That is silly." And that is my point. Yet this makes as much or more sense than many things that I've seen in the Secular Humanist-dominated textbooks. In fact, there is as much if not more evidence backing up this hypothesis as there is in favor of the evolutionary trees-of-life that adorn textbooks around the globe today.

Think it through and be prepared to answer the next person who claims that we evolved from apes. Just reply, "Why not snow cones?"

"I wish I were younger. What inclines me now to think you may be right in regarding [evolution] as the central and radical lie in the whole web of falsehood that now governs our lives is not so much your arguments against it as the fanatical and twisted attitudes of its defenders."

C.S. Lewis, in a letter to Capt. Bernard Acworth of the Evolution Protest Movement, 1951

FIFTEEN

Darwinists' Dating Methods Are Tools of Deception

Darwinian-biased scientists use the ultimate dating game in their efforts to convince billions of people that we live on a planet that is *billions of years* old, and to support their claims that fossils of humans have been found that are over a *million years* old, plus fossils of various creatures that are even much older than that.

In this chapter we will focus on the Potassium-Argon and Carbon-14 dating methods, and see how they stand up under the microscope of observable science. We will even show how our world's coal deposits and every woman's diamond engagement ring is proof of a young earth that is only thousands, and not *billions*, of years old.

Is Potassium-Argon Dating Reliable?

When compared to real, observable science, radiometric (also called radioisotope) dating methods used by old-earth-believing scientists fall far short of being empirical science. Many wild guesses, referred to as *assumptions*, are made which corrupt the reliability of the radioisotope dating techniques.

Bear in mind that these radioisotope dating methods are not a measure of time, but rather a method of accounting for the amount of radioactive decay that has occurred within a certain type of radioactive element.

Let's consider Potassium-Argon (K-Ar) dating.

It is a scientific fact that Potassium 40 decays into Argon 40. It is also a scientific fact that a scientist trained to work with radiometric dating techniques can grind up a rock and very accurately measure the amount of potassium 40, and the amount of Argon 40, that is contained in the rock sample that he or she is working with.

These are plain and simple scientific facts that can be tested, observed, studied and repeated. It is also a plain and simple fact that this is where the science comes to an end with regard to the radiometric dating methods.

Now, to derive the age of the rock based on the amount of radioactive decay which has occurred, several wild guesses – those *assumptions* again – must be made. It is these non-observed assumptions that completely corrupt the integrity of the ages obtained by the various radiometric dating methods.

The following examples point out major faults with Potassium-Argon dating:

Example Number One:

A scientist using the radiometric dating methods to ascertain the age of a rock must assume that the rate of decay from Potassium 40 to Argon 40 has always remained the same. [The notion that something observed today has always remained basically the same, such as the uniform rate-of-decay, is called Uniformitarianism.] Since no one was there to test, study or observe that the rate of decay has always been the same, this single wild guess, if incorrect, will cause the supposed age obtained by this particular dating method to be off by *millions or billions of years.*

Example Number Two:

A scientist using the K-Ar radiometric dating method in order to derive the age of a rock must assume that there was not any Argon 40 in the rock when the rock first formed. Next, by measuring the amount of Argon 40 currently in the rock, combined with the assumption that the formation rate of Argon 40 has always been the same, an age is assigned to the rock. Since no one was there to test, study or observe that there was not any Argon 40 in the rock when the rock first formed, this single assumption, if incorrect, will cause the supposed age obtained by this particular dating method to be off by *millions or billions of years.*

Example Number Three:

A scientist using the Potassium-Argon radiometric dating method in order to assign an age to a rock must assume that the rock was never contaminated with either Potassium 40 or with Argon 40. In other words, everyone is supposed to believe that this rock laid there for *millions or billions of years* and was never contaminated by the addition or loss of the elements being measured. Since Argon 40 is a gas that can easily pass from one rock to the next this is an incredulous assumption, especially since it is well documented that heat, pressure, earthquakes and moisture can all cause contamination to occur. Yet we are to blindly trust that the rock never lost or gained either Potassium 40 or Argon 40.

As in the previous two examples, since no one was there to test, study or observe that the rock was never contaminated with either Potassium 40 or with Argon 40, this single wild guess, if incorrect, will cause the supposed age obtained by this particular dating method to be off by *millions or billions of years.*

These three wild guesses apply to all of the various radiometric dating methods.

And these three assumptions alone reveal the corruption involved in the radioisotope dating techniques. These methods are completely unreliable when used to obtain the age of any particular rock, fossil or strata layer.

Is Carbon Dating reliable?

As is the case with radiometric dating methods, Carbon Dating is not a reliable means of verifying that the earth is even thousands of years old, much less *millions or billions of years* old as long claimed by Naturalistic-believing scientists.

Both radiometric dating methods and Carbon Dating rely on unfounded assumptions which corrupt the reliability of the dates that these methods obtain.

Carbon Dating relies on several assumptions which corrupt the reliability of the dates that it obtains. As these erroneous assumptions are extrapolated backwards over long ages of imagined time, there is a multiplying effect on how these wild guesses throw off the dates being attained. Many honest scientists do think that Carbon Dating may be fairly accurate when used to date items less than 3,000 years old. This is due to the fact that there hasn't been enough time for the false assumptions to completely corrupt the dates that are obtained. However, any accurate Carbon dates of things less than 3,000 years old are most likely due to "calibrating" the dates to fit with known historical events. Thus, specific dates obtained from the Carbon Dating method which are not substantiated by historical events are not likely to be reliable.

I will get to the unfounded assumptions used in Carbon Dating in a moment.

Realize that by looking at abundant and observable strata and fossil evidence from around Planet Earth, observable science verifies two points. First, specific dates are not likely to be valid. Second, that Carbon Dating actually supports a young earth viewpoint, which is that the earth's strata layers can only be a few thousand years old and were laid down during a single global flood.

How can this be?

The following discussion of Carbon Dating will demonstrate the validity of what the observable evidence indicates.

In Carbon Dating the amount of Carbon-14 in organic remains is measured. Carbon-14 is produced in the earth's atmosphere and during the process of photosynthesis plants take in CO_2, which contains trace amounts of Carbon-14. When an animal breathes in air, or eats a plant, it also takes in trace amounts of Carbon-14. Once the plant or animal dies it stops ingesting additional amounts of Carbon-14 and, since Carbon-14 decays away over time, the less Carbon-14 found in an item the older will be the age assigned to that particular remain.

To derive the age of an item based on Carbon Dating, several wild guesses, which are referred to as assumptions, must be made. It is these assumptions that completely corrupt the integrity of the Carbon Dating method.

For example, a scientist using the Carbon Dating method to ascertain the age of an organic remain must assume that the rate of Carbon-14 decay has always remained the same. Since no one was there to test, study or observe that the rate of decay has always been the same, this single wild guess, if incorrect, will cause the supposed age obtained by this particular dating method to be completely in error.

Furthermore, scientists seem to agree that measurable amounts of Carbon-14 should decay away in less than 100,000 years. Many scientists claim the timeframe may be closer to 50,000 years but to be both generous and fair, let us use the longer duration of time. Therefore, Carbon Dating cannot age an item older than 100,000 years since there would be no measurable Carbon-14 left if the item being dated is actually that old, or older.

This is a point of particular interest because secular teachings claim that the Cambrian layer, the lowest stratified rock layer with appreciable amounts of fossils in it, is supposed to be up to 580,000,000 years old, which is 5,800 one-hundred-thousand-year periods. Obviously, if the secular teachings were true no Carbon-14 would remain in the Cambrian layer today, yet scientific studies have revealed that all fossil-bearing layers, down to the supposedly 580-million-year-old Cambrian strata layer, still contain measurable amounts of Carbon-14. This by itself proves that the earth's strata, from which all old-earth beliefs have been derived, can only be a few thousand years old!

Of even more interest is that the range of amounts of Carbon-14 found in all the fossil-bearing strata is in the same range of amounts from the top layers all the way down to the bottom layers. Since Carbon-14 decays away over time, this clearly demonstrates that all the earth's stratified layers were formed during a single event. Only a global flood can viably account for this.

That is correct:

The testable scientific evidence supports a relatively young earth which has endured a global flood!

So Carbon Dating actually supports the Biblical account that the earth was judged by a global flood in the recent past. By the recent past I mean thousands of years as compared to the *millions of years* that are required by various non-Scriptural old-earth beliefs from Theistic Evolutionism to Progressive Creationism to all Naturalistic and Darwinian belief systems, which do not allow for any worldwide flood of Biblical proportions.

This is prophesied in 2 Peter 3:3-6, a passage we quoted previously in this report that is worth repeating:

> **[3]** Knowing this first, that there shall come in the last days scoffers, walking after their own lusts,
>
> **[4]** And saying, Where is the promise of his coming? for since the fathers fell asleep, all things continue as they were from the beginning of the creation.
>
> **[5]** For this they willingly are ignorant of, that by the word of God the heavens were of old, and the earth standing out of the water and in the water:
>
> **[6]** Whereby the world that then was, being overflowed with water, perished:

Although the Bible is not a book of science, it is the true history book of the universe and Scripture discusses nothing that conflicts with real, testable science. This, I believe, is sufficient reason to consider that what the Bible says about a flood that covered all the high hills under heaven and destroyed every living thing on earth should be taken into consideration by scientists sincerely seeking the truth about our world's geology. If the Bible's account of a global flood is true it would mean that today's earth and its geology is drastically different from the original creation.

Look at it this way.

A global flood would have buried huge quantities of living plants and animals in sedimentary layers. This fact is observable in the form of fossil fuels such as coal, natural gas and oil discovered in the earth's stratified rock layers. The amount of coal layers alone indicate that the earth, as originally created, was covered with tremendously lush vegetation as compared to the post-flood world we live in today. Vast quantities of carbon were also buried with those plants and animals.

What this means in terms of Carbon Dating plants and animals from the pre-flood world, those found buried in the earth's strata layers today, is that they would have had far lower ratios of Carbon-14 to Carbon-12 than do post-flood organisms. In fact, some estimates run as high as pre-flood animals or plants having contained less than 2% of the Carbon-14/Carbon-12 ratio that is observed in the post-flood world today.

Since it is the ratio of the Carbon-14 found in remains as compared to the Carbon-12 observed in the atmosphere today that is used to obtain ages via the Carbon Dating method, the pre-flood entities would have started out with very little Carbon-14 and will date much older than they actually are.

Another issue for the reliability of ages obtained via Carbon Dating is the scientifically observed fact that the earth's magnetic field has weakened by 6% over the past 150 years. This means that the field was stronger in the past and would have blocked more of the rays from the sun. This fact alone would have reduced the amount of Carbon-14 forming in the atmosphere and lowered the Carbon-14 amounts found in flood-deposited strata.

However, Carbon-14 dating continues to be used without any regard to the possibility that a global flood occurred or that conditions that lent themselves to the formation and collection of Carbon-14 might have changed dramatically. Either atmospheric changes or a worldwide flood, or as is most likely, both, would make ages derived from Carbon Dating inaccurate.

The religious—not scientific—bias against God's Word leaves old-earth worshipping secular scientists and scientists who have been honestly fooled into thinking that secular beliefs are *science* caught up in a real quandary.

The quandary is this: Do they follow the facts to wherever they lead or do they force-fit the evidences to blend with secular beliefs?

For one example, secular teachings claim that the Carboniferous layer formed about 250,000,000 years ago. As we have mentioned, scientists generally agree that measurable amounts of Carbon-14 should decay away in less than 100,000 years, and 250-million years is 2,500 one-hundred-thousand-year periods of time. Therefore, Carbon-14 should never be found in the Carboniferous layer, unless of course the layer is less than 100,000 years old.

Here is the quandary for the old-earth believers.

Scientific studies on coal recovered from the Carboniferous layer reveal that all coal deposits still contain measurable amounts of Carbon-14. In fact, never has any coal layer been found that does not contain measurable amounts of Carbon-14. By itself, this is strong proof that the earth's strata layers, from which all old-earth beliefs have been derived, can only be a few thousand years old, and the only viable explanation as to how the earth's strata layers all formed recently and during the same event is that the world we live on endured a global flood that covered all of the high hills under heaven sometime within the last few thousand years.

The modern-day scientist is thus left in a huge quandary: either admit that the old-earth beliefs have no viable basis because the earth's strata layers, from which the old-earth beliefs have been derived, formed recently during a worldwide flood, or toe the old-earth line. To do the honest thing would be to commit career suicide as *millions of years of time* is the magic ingredient for Darwinism, and Darwinism provides the foundation for Naturalistic-worshipping Secular Humanists. This is why most researchers choose the latter and become a religious zealot as opposed to a fact-following scientist.

Diamonds prove our planet is young, and not *billions of years* old.

Because Carbon Dating is so often referred to, it is important to continue the discussion about how Carbon Dating actually supports the viewpoint that we live on a planet which is thousands of years old, rather than on a planet that is four-plus *billion years* old as Naturalists, Darwinists and their allies claim that it is.

Secular teachings claim that most diamonds are a billion or more years old, which is at least 10,000 one-hundred-thousand-year periods of time. Therefore, if diamonds are indeed *billions of years* old, these romantic sparklers should never contain Carbon-14.

However, the Carbon Dating of diamonds from around the world provides strong proof that the earth is only a few thousand years old. Recent studies of diamonds obtained from various diamond mines around the globe reveal that the sparklers still contain measurable amounts of Carbon-14. This proves that the diamonds can only be a few thousand years old, not *millions*, much less *billions*, of years in age.

Again, Carbon-14 Dating is proving to be unfriendly to the old-earth claims, which are a prerequisite for Darwinism and the

Secular worldview which is based upon *billions of years leading to Darwinian-style evolution* being true.

Further proof that the various radiometric dating techniques do not reveal reliable ages for fossils, rocks or strata layers is seen by the observable fact that rocks sent to laboratories to be dated by various radioisotope methods vary tremendously in the ages obtained. Whether using the same particular isotope method at different labs, or different methods at the same lab, ages obtained from any particular item can vary by *billions of years.*

This from the *Anthropological Journal of Canada* (R Lee; Radiocarbon, Ages in Error; Vol 19, No 3, 1981, pages 9, 29.):

"The troubles of the radiocarbon dating method are undeniably deep and serious...half of the dates are rejected...there are gross discrepancies...accepted dates are <u>selected dates</u>."

Selected dates?

You mean that secular scientists pick a date that fits their unobserved belief in the slow, gradual formation of the sedimentary layers that were laid down by water (as opposed to accepting that the layers were laid down quickly and recently during a global flood)? Exactly.

And where do they *select* their dates from? From the man-made Geologic Column or Time Scale.

"Time is in fact the hero of the plot...the impossible becomes possible...time itself performs the miracles."

George Wald, famous Darwinist who was a Harvard professor and Nobel Prize winner

SIXTEEN

Worshipping at the altar of time

Darwinists indeed worship time. They also worship a Geologic Column that is non-existent in the natural world.

Worship is the correct word because Darwinism is a religion.

I have seen few people with faith as strong as Secular Humanists have. I actually admire their conviction to the religious philosophy of naturalistic evolutionism. However, I certainly do not admire the fact that they sell their religion to innocent young minds in the public schools with the use of out-and-out frauds as we prove in this report. I wish that more Christians would be as loyal to the Word of God as Secular Humanists are to the theory of Darwinism.

**So why do Darwinists deem
it necessary to worship time?**

The inescapable truth is that time is their magic ingredient. It is so revealing what former Harvard professor and Nobel Prize winner George Wald stated:

> **"Time is in fact the hero of the plot...the impossible becomes possible...time itself performs the miracles."**

Yet the old-earth ages required by Darwinists come from just two primary sources: the Geologic Time Scale (also referred to as the Geologic Column) and the radiometric dating methods.

In the previous chapter I refuted the radioisotope dating methods, revealing that they get a wide range of ages from any particular rock. These methods are based on selecting a date that matches the Geologic Time Scale. In other words, the published dates come from this man-made time scale, and not from any scientifically valid method of dating things.

However, the lack of real scientific support has never deterred or even slowed down the Darwinian faithful.

Evidence is overwhelming that the Geologic Column is interpreted through a bias against Christianity, Biblical accounts of Creation and the global flood.

The Geologic Column was popularized in the early 1800's. Many Christian geologists and scientists from other fields of study were involved early on, when the Geologic Column was developed as a base from which to conduct study of the earth's rock layers. However, it did not take long for Secularists to undermine the actual science and employ the Column as a time scale that would undermine billions of people's faith in the Word of God. Each of the sedimentary layers of rock, which were laid down by water, were soon given a name, an ancient age, and assigned corresponding index fossils.

However, the scientific problems with the Geologic Column, especially when used as a scale of time, are many. First was the anti-Christian bias that permeated the ages assigned to the layers.

Despite the fact that the layers were laid down by water, complete denial of the global flood was invoked and ancient ages were given to the layers with the purpose of undermining the Bible's accounts of world history.

**And where did these secular scientists
derive the ages of the layers?**

Keep in mind that this was about the time that George Washington died...a time when the latest scientific thought was that a living cell was just a glob of gelatin-like substance. So again I ask, "Where did the secular scientists derive the ages of the layers?"

They made them up. Pure and simple.

Then they assigned index fossils to each layer. Index fossils are any fossilized remains of a plant or animal that supposedly went extinct while that particular layer was slowly developing. Since that creature supposedly went extinct while that layer was forming, it should never be found above that particular strata layer. By assigning an age to an index fossil, any strata layer found containing that index fossil, and everything in that rock layer, is given the age assigned to that particular index fossil.

Remember that secularists believe the layers formed over long periods of time so the deeper layers are presumed to be *millions of years* older than the layers above them.

The problems with the Geologic Time Scale are numerous, and keep in mind that this is where the old-earth dates are actually derived from. For instance, some of the index fossils, supposedly extinct for up to *hundreds of millions of years*, keep showing up alive today. This alone refutes the time scale, but the situation gets even worse from a scientific standpoint.

The Geologic Column is comprised of twelve primary strata layers; however, these twelve layers, with their corresponding index fossils by which they date the rock layers, have never been found anywhere in the natural world in the order depicted by the Geologic Time Scale.

Never.

And again, this time scale, non-existent in the natural world, is the primary source that Secularists use to ascertain the age of the earth.

The only places that the Geologic Column has been found in its entirety, including being in the proper order, are in school textbooks and in museum depictions. Most of the earth averages only three or four of the layers and these are generally in a mixed order, often with the supposedly older layers positioned on top of the supposedly younger layers.

So with all due respect to Dr. Wald, and all those who worship at the altar of various naturalistic philosophies or old-earth beliefs, real science shows that there is no viable reason to believe in an old earth. The fact is that time itself did not and could not perform the miracles required by the theory of Darwinism.

We discuss at length how and why geological evidence points to a young earth in Report No. 3: **371 Days That Scarred Our Planet: What the stones and bones reveal might surprise you.**

"Evolutionism is a fairy tale for grown-ups. This theory has helped nothing in the progress of science. It is useless."

Professor Louis Bounoure, Past President of the Biological Society of Strassbourg, Director of the Strassbourg Zoological Museum, Director of Research at the French National Center of Scientific Research. (Quoted in *The Advocate*, March 8, 1984.)

SEVENTEEN

The Great Dinosaur Scam

According to my co-author Jim Dobkins a newspaper editor once told him, "Never let the facts interfere with a good story."

The reason I bring this up is because that is exactly the motto that Darwinian-flag-waving Secular Humanists live by. We've discussed how good they are at drawing pictures of creatures that never existed in order to support their theory that never took place. Well, they're just as creative in spinning yarns about dinosaurs. In fact, their claims about dinosaurs having gone extinct *millions of years ago* have fooled billions of people into believing in the magic ingredient needed for Darwinian-style evolution to seem possible—*millions of years* of time.

I can sum up my response to their claim about dinosaurs in one word:

Hogwash!

As we discussed in a previous chapter, Charles Darwin believed that if his theory were true, numberless intermediate species, transitional links that is, had to have existed. Yet today, not a single such transitional kind is known that will stand up to honest scientific scrutiny. Not a single one from the fossil record (Darwinists claim several candidates; however, these do not stand up to honest evaluation) and none are found from among the millions of living species alive on earth today. Not a single one. They are all missing and this is why they are called the *missing links*.

Dinosaurs are a good example of the gaping hole encountered by Darwinists due to these transitional links not existing. Although most people think that all dinosaurs were giants, and several were, the actual size of the average dinosaur was the size of a large sheep. In fact, several were only about the size of a chicken.

One of the problems that dinosaurs present to Darwinian fundamentalists is that they show no evidence in support of Darwin's theory. The huge beasts appear suddenly in the fossil record. There is no evidence that they evolved from anything. We are supposed to believe that dinosaurs, such as an eighty-ton sauropod, evolved from something but left no traces of what they evolved from. It is certainly reasonable and fair to expect those espousing Darwinism as if it were science to actually provide some scientific proof in support of their claims. Yet dinosaurs, like everything else, fail to provide any evidence in support of Darwinian teachings.

None.

Next, during the supposed 190-million-year reign of these incredible creatures, the fossil record has not revealed any evidence that they were evolving amongst themselves. They are found fully developed, and nothing is found that supports that they were evolving during all those many imagined *millions and millions of years* of time.

According to secular interpretations of the earth's supposed gradual formation of its rock layers, several dinosaurs went extinct before others did. The ages are based on which water-borne sedimentary strata layer they are found in – denying that the layers were laid down rapidly during the global flood.

But going extinct is not evolving, and no evidence of dinosaur evolution is found in ANY sedimentary strata layer.

Finally, dinosaurs disappear quite suddenly, revealing nothing of their evolutionary change into anything else. I see a clear pattern here in the theory of Darwinism. They have no evidence to back up their claims.

None.

Dinosaurs do not show any traces of having evolved from something else. Dinosaurs reveal no evidence of evolving into anything else. These oftentimes incredible beasts show no evidence in support of Darwinism. Not a single example.

Actually, dinosaurs are no different than any other animal in that they provide no evidence in support of Darwin's theory.

None.

And no, it is not because of "punctuated equilibrium," the Darwinian fundamentalist's excuse that evolution happened too fast to leave any evidence behind.

And, as usual, the fossil record is a total embarrassment to Darwinism.

Evidence indicates dinosaurs lived at the same time as humans.

Secular interpretations of the evidence continue to expound that dinosaurs lived about a quarter of a *billion years ago* and that they went extinct more than 64 *million years ago*. Their extinction, according to secular beliefs, occurred about 63 *million years* before humans came along.

If these incredible beasts have been extinct for *millions of years* we should not find any evidence in support of their existence during the past million years, much less any facts showing that they lived within the past 6,000 years. If the secular teachings were true we should not find any evidence of man having lived at the same time as dinosaurs.

None. Zip. Nada.

Yet historical accounts and recently-unearthed evidence bear witness that dinosaurs lived at the same time as humans. In fact, human historical records, archaeology and other scientific finds continually run afoul of these Darwinian claims.

Note that the word dinosaur was not proposed until 1841 by Sir Richard Owen (1804-1892), a pioneering British comparative anatomist. Owen coined the term dinosauria (from the Greek "deinos" meaning fearfully great, and "sauros" meaning lizard). Prior to 1841 they were called dragons and/or serpents. And ancient history books are full of thousands of accounts of men and various kinds of dragons. We call these *dragon stories* today.

Consider three such accounts that came out of what is now India:

> 2,300 years ago Alexander The Great wrote that his soldiers were scared by great dragons that lived in caves there.

> 2,000 years ago Roman historian Pliny the Elder wrote that the "...elephants are constantly at war with the dragons" there.

> Then 1,900 years ago Apollonius of Tyana recorded that "the whole of India is girt with enormous dragons...killers of elephants."

Well, it takes a pretty big critter to kill and eat an elephant. In fact, Marco Polo wrote of the dragons that were domesticated in China just 750 years ago. There was nothing mythical about dragons, which were actually dinosaurs, a few hundred years ago.

Paintings and carvings of various sorts of dinosaurs in and on rock walls, etched stones and artwork are found around the globe. We are told that many of these man-made depictions are from 800 to 2,000 years old. Yet we only officially recognized dinosaur bones in 1821 (although it would be 20 years later before they were actually called dinosaur bones).

Native Americans of the southwestern United States told accounts of what was called the *Thunderbird*. Carvings of this creature look like a Pterodactyl-type flier. These supposedly went extinct with the dinosaurs.

Doesn't it make sense that people had to have seen dinosaurs in order to replicate them in carvings and paintings 1,800 years before mankind began to acknowledge their fossils?

It makes perfect sense to me.

What doesn't fit with the facts is the tale that dinosaurs have been extinct for *millions of years* before humans existed.

True science continues to make things worse for Darwinism. Over the past twenty years scientists have found several non-fossilized dinosaur bones that still contain red blood cells and soft, flexible tissues inside of the bones. Researchers admit that these things could not have lasted for more than 10,000 or so years. They most certainly are not *millions of years* old.

Time beyond human comprehension

Darwinism has to have time beyond human comprehension to fool people into believing that humans evolved on their own without God being involved, and dinosaurs are one of their cornerstones to getting you to believe in their much-needed *millions of years* of time.

However, ask yourself this simple question:

Who saw dinosaurs go extinct 64 million years ago?

Then look at the actual evidence and you will discover that dinosaurs are not a friend to Darwin's theory of evolution.

Dinosaurs are mentioned and described in the Bible.

If the Bible is true – and I believe it is true word for word and cover to cover – then dinosaurs indeed lived at the same time as humans. In fact, we were both made on the sixth day of Creation.

Numerous passages in the Bible refer to dragons and serpents, although they are not described in great detail. However, here's a passage from the Book of Job that many Bible students and scholars clearly consider to be a description of a large dinosaur. In this case the word behemoth is used as opposed to the term dragon.

Job 40: 15-24

[15] Behold now behemoth, which I made with thee; he eateth grass as an ox.
[16] Lo now, his strength is in his loins, and his force is in the navel of his belly.

[17] He moveth his tail like a cedar: the sinews of
his stones are wrapped together.
[18] His bones are as strong pieces of brass; his
bones are like bars of iron.
[19] He is the chief of the ways of God: he that
made him can make his sword to approach unto
him.
[20] Surely the mountains bring him forth food,
where all the beasts of the field play.
[21] He lieth under the shady trees, in the covert of
the reed, and fens.
[22] The shady trees cover him with their shadow;
the willows of the brook compass him about.
[23] Behold, he drinketh up a river, and hasteth not:
he trusteth that he can draw up Jordan into his
mouth.
[24] He taketh it with his eyes: his nose pierceth
through snares.

Job is believed to have lived around or possibly a bit before
B.C. 1,500, at least several hundred years before the time of
Moses. He lived in the land of Uz, which is mentioned several
times in the Old Testament.

In this passage God is speaking to Job in what is widely
considered one of the loftiest examples of classic literature. For
Job to clearly understand what God is talking about in discussing
His creation of behemoth he would have to be aware of the
existence of this great creature.

Strength in the loins was required in order to balance the long
and heavy tail and neck. In fact the tail is compared to a cedar tree
and its bones likened to bars of iron. This sounds like the
description of a saurapod-type dinosaur to me.

I find it sad that theologians who have been fooled into
believing in *millions of years of time* try to explain away behemoth
by suggesting that the Bible passage is describing an elephant or
hippopotamus. But have you looked at the tails of these animals?

An elephant's tail can aptly be described as looking like a dangling whip while a hippo's tail looks more like a short stump. Neither bears any resemblance to the long trunk of a large cedar tree.

I think that the dinosaurs were indeed the "chief of the ways of God" and that God deserves the credit for His created beings. I also think it is time that Bible believers stand up and reclaim dinosaurs for the glory of their Biblical Creator as opposed to continuing to allow Satan to use them to undermine the faith of billions of unsuspecting people around the world.

So what caused dinosaurs to become extinct?

There are a thousand theories as to what caused dinosaurs to suddenly disappear. Again, I will point out that we now find their bones buried in sedimentary layers which were laid down by water.

It appears rather obvious that dinosaur fossils were buried in sedimentary layers of strata that were laid down by water during the Biblical global flood judgment of man's sin. This is the only viable account that fits with the observable evidence. We dig deeply into the proof of the worldwide flood in Report No. 3: **371 Days That Scarred Our Planet: What the stones and bones reveal might surprise you.**

Yet Secular Humanists, who control the educational and scientific establishments, are persistent in not letting the facts interfere with a good story.

"So heated is the debate that one Darwinian says there are times when he thinks about going into a field with more intellectual honesty: the used-car business."

Sharon Begley, "Science Contra Darwin," *Newsweek*, **April 8, 1985, p. 80.**

EIGHTEEN

Solar Winds
and
Life on Mars

The concept that there is life on other planets and in other galaxies is widely popular. It is the core idea of countless science fiction novels. Many television shows, including the classic Star Trek series, and major movies have been spawned by this concept.

Some blockbuster movies that come to mind include Close Encounters of the Third Kind, 1977; Star Wars, 1977; E.T. The Extra-Terrestrial, 1982; Independence Day, 1996; and Signs, 2002.

Moviegoers find such film-fare fascinating. They want to believe that life on other planets and in other galaxies is possible.

Meanwhile, Secular Humanists continue to flaunt the flag of Darwinian-style evolution, claiming that life will be found on other planets. In support of this contention the United States government is spending billions of taxpayer dollars searching for life in outer space.

However, having discovered the complete impossibility of life having begun spontaneously from non-living matter here on earth, and completely closed to the fact that life might have been created by God just as the Holy Bible claims it was, they are simply sending their problem to where no one other than their own institutions can conduct the observations.

Thus the world will be subject to their interpretations of whatever they find. Actually, this is not unlike the way the people of the United States have been subjected to the Humanistic interpretations and outright frauds employed to fool folks into believing in Darwinian evolution for many decades. However, moving the issue to where only institutes with their own spacecraft can conduct the research will make exposing any frauds next to impossible. This is not how real science functions.

Of course as we have previously pointed out, Darwinism is not science; it is a foundational belief for the religious worldview of Secular Humanism. Consider the current, aggressive search for life on Mars, for example.

Here's where those solar winds come into play.

It is an observed scientific fact that micro-organisms from earth waft up on wind currents. These have been found in the earth's upper atmosphere and are assumed to get pushed out into space by the solar winds emanating from the sun.

These solar winds push things away from the sun and since Mars is further away from the sun than the earth the theory is that whenever the microbes get within range of Mars' gravitational pull, they will be brought down onto the surface of the planet.

This is what NASA's Mars landings and the search for life there are truly all about. The Secular Humanist crowd, devoid of any real scientific evidence in support of *from-goo-to-you* Darwinian-style evolution on our planet, hope to recover some of these earth-borne microbes from the surface of the Red Planet.

Then headlines around the world will scream:

LIFE EVOLVED ON MARS!!!

Keep this in mind...because if it happens, as I suspect that it one day will, the discovery will be used to mislead billions of people. Such deception may turn out to be one of the greatest frauds perpetrated in the name of Secular Humanism in the history of the world. And that is saying quite a lot.

Mark these words, should this hoax come to fruition, the proof that the microbes came from earth will be that they will be like the microbes we find here today. This will be the case due to the microbes having originated on earth before getting wafted up by air currents and pushed to outer space by the solar winds.

"I have come to the conclusion that Darwinism is not a testable scientific theory, but a metaphysical research programme..."

Dr. Karl Popper, German-born philosopher of science, called by Nobel Prize-winner Peter Medawar, "incomparably the greatest philosopher of science who has ever lived."

NINETEEN

It Must Be Refutable

This is the chapter alluded to at the end of Chapter Five (our discussion about Intelligent Design and Intelligent Biblical Design) in which we will compare how Intelligent Design, Darwinism, and Biblical Creationism stand up under the microscope of real science.

Modern-day public schools and colleges teach that Darwinism is science. However, Darwinian evolution is a religious belief about how we came about and it provides the philosophical foundation for Secular Humanism.

In other words, Darwinism is today's foundation for Secular Humanism, a worldview which is basically materialistic atheism, and it is important for everyone to understand this so they will not be fooled into thinking that Darwinism is science.

Real science is knowledge which has been derived from the study of testable and repeatable events. A true scientific hypothesis must be:

A. Predictable and

B. Refutable.

This means that the hypothesis must:

A. Make testable predictions that are expected to be found to be true should the proposed hypothesis be accurate and

B. Be possible to be proven wrong. For instance, the hypothesis would be refuted if its predictions were not found to be accurate.

Let's look at the predictions and refutability of Naturalistic Darwinism which is the philosophical foundation for Secular Humanism.

Prediction Number One:

Darwinism predicts that biology will find organisms adding new and beneficial genetic information to existing gene pools. This would be reliable, empirical criterion of Darwinism.

And what does observable science find?

Well, real science has never discovered a single method by which nature can add appreciable amounts of new and beneficial genetic information to a plant or animal's gene pool.

Increasing genetic information has proven to be another major hurdle for Darwinism.

Although Secularists have been teaching Neo-Darwinism, that mutations are what add the needed new and beneficial genetic information which then leads to Darwinian-style change, millions of scientific tests prove otherwise. Scientific research proves conclusively that mutations are caused by the mix-up or loss of their inherited gene pool so they are usually genetically weaker than their parent form.

Because of this, in a natural environment, mutations tend to be eliminated by Natural Selection. With regard to the first prediction, Darwinism fails this key criterion.

Prediction Number Two:

Darwinism predicts that geology will find the fossil record to be filled with millions of transitional kinds, one species evolving into another, of both plants and animals. These would be reliable, empirical criteria of Darwinism.

But what does observable science find?

The overall fossil record is an embarrassment to Darwinism. All higher kinds of plants and animals appear abruptly in the fossil record with zero transitional types linking one group to another. The amount of transitional forms involved in the evolutionary transformation from a single-celled creature to every life form found on earth today, and to the now extinct forms found only in the fossil record, should total an incredible number.

Yet the only fossils presented as candidates for missing links have not held up to even scant scientific scrutiny. In fact many—such as the infamous Piltdown Man—have been proven to be frauds which have been used to mislead billions of people into believing in Darwinian-style evolution. However, with regard to observable science and the second prediction, Darwinism again fails its own key criterion.

So what about the refutability of Naturalistic Darwinism?

Because Humanists dominate the educational and the scientific establishments we must look past what they say and study the actual facts. Humanists argue that unless every conceivable possibility to support materialism can be proven impossible, no matter how improbable (such as aliens may have dropped us off), only their philosophy can be considered as a possible reason for the origins of life. Secular Humanists force all opposing views to prove a negative, which is impossible.

Although Darwinism goes against multiple scientific laws and principles, and despite the fact that not a shred of evidence exists that Darwinian-style change has ever taken place, Humanists have made it impossible to refute Darwinism unless you can examine the entire universe and show that their religious belief never took place anywhere. This makes Naturalistic Darwinism a non-refutable religious dogma, and not science.

Darwinian Evolution is a religious dogma which fails its own critical criteria and is non-refutable. However, the failed predictions of Darwinism refute Secular Humanistic teachings that we all evolved on our own.

Sadly, only the secular interpretations are shown to public school and college students today.

Is Intelligent Design predictable and refutable?

Intelligent Design's predictions:

ID predicts that biology will find organisms with both Specified Complexity and Irreducible Complexity. Logic holds that an Intelligent Designer would most likely leave behind his mark, sort of like your touch leaves behind a fingerprint. Both Specified and Irreducible Complexity are reliable, empirical criteria for ID.

So what is Specified Complexity?

If something shows that it accomplishes a specific function which is too complex to be repeated by random chance then it shows specified complexity that materialistic means alone cannot account for.

For example, if I shot an arrow from a bow at a target and hit the bulls-eye once, well, random chance could account for that. However, if I did so over and over again, you could say that the result was due to an intelligent source being behind the arrows continually hitting their target. This event would show specified complexity.

And what is Irreducible Complexity?

Well, whenever a functioning system of interrelated parts, like a car, has a minimum amount of parts without which the system will fail to function, then the system exhibits irreducible complexity. That minimum number of parts must be there from the start or that system never would have begun to operate.

For example, the irreducible complexity of your immune system shows specified complexity.

As you read this report, various bacteria and viruses are trying to devour you and they would succeed if your immune system were not fully operational. Our immune systems are *Irreducibly Complex*. They consist of billions of bone marrow cells, antibodies and proteins.

Antibodies identify invaders from normal body cells but can only bind to one kind of virus or bacteria. Hence there are billions of different antibodies, each made by its own cell in the bone marrow. When the specific invader shows up, that antibody must inform its cell to mass produce more of that antibody. To do this the antibody remains attached to its producing cell.

Once bonded to the invader, a piece of the invader is attached to a specific protein molecule which attaches itself to a helper cell which in turn attaches to yet another type of protein.

On and on this goes until the production cell gets the message to mass-produce its particular antibody. Then the cell produces specialized plasma cells which are not attached to the cell so they can travel through your bloodstream and hunt down the invader. If any of the bone marrow cells, antibodies or proteins were to fail to perform their function when their specific invader appeared the entire system would fail and we would die. Just as are most of the organs and systems in our bodies, our immune systems are irreducibly complex.

The Immune system exhibits both Specified and Irreducible Complexity. The system was designed to perform the specific function of protecting us from invaders. Every part needed to be there from the start or the immune system would fail.

Intelligent Design meets these key criteria.

With regard to being scientifically refutable, if it could be proven that materialistic means alone can produce complex biological systems, ID would be refuted.

Many branches of modern science, such as Forensics and Archaeology, already employ Intelligent Design. When an Archaeologist finds a stone arrowhead how does he know it was not the product of rain erosion? He knows because the sculpted stone reveals too much purposeful ID to have formed on its own.

Another fact is that the only known mechanism for causing various parts to assemble and form a complex machine, like an airplane, is through human engineering, which is an example of Intelligent Design.

ID meets its own key criteria, is able to be scientifically refuted, and is already used in various fields of research. These facts alone should qualify ID as a viable branch of scientific study. Only Humanistic bias, masquerading as science, is preventing science from moving ahead. This is yet another example of how Naturalistic beliefs are undermining real scientific research.

Is Biblical Creation scientifically sound?

Modern-day public schools and colleges teach that Biblical Creation is a religious belief – which it is, just as is Darwinian evolution.

Both are philosophies on how life originated and came to where we are today. Biblical Creation also provides the foundation for the Gospel of Jesus Christ while Darwinism provides the foundation for Secular Humanism.

And how does Biblical Creation stand up to scientific facts?

Remember, real science is knowledge which has been derived from the study of testable and repeatable events. A true scientific hypothesis must be:

A. Predictable and

B. Refutable.

This means that the hypothesis must:

A. Make testable predictions that are expected to be found to be true should the proposed hypothesis be accurate and

B. Be possible to be proven wrong. For instance, the hypothesis would be refuted if its predictions were not found to be accurate.

Now let's look at the predictions and refutability of Biblical Creation.

Ten times in the Book of Genesis God's Word predicts that plants and/or animals will "bring forth after their kind."

Also in Genesis and throughout the Bible, Scripture claims that the world endured a global flood where everything living on the earth died and the highest hills were covered by the water. In 2 Peter 3:6, we are told that "Whereby the world that then was, being overflowed with water, perished…"

Two key predictions of Biblical Creation:

Thus Biblical Creationism predicts that biology will find that plants and animals will only reproduce their own kind with variations occurring within their kind; and that geology will find that the earth has endured a worldwide, catastrophic, aqueous event. These would be reliable, empirical criteria of Biblical Creation.

Prediction Number One:

Biblical Creationism predicts that biology will find that plants and animals will only reproduce their own kind of plant or animal with adaptational changes only occurring within that particular kind of being.

Micro-evolution (better referred to as micro-adaptations because most people associate the word evolution with the Darwinian-style change of one kind of being into a different kind of creature) is a scientific fact. Millions of examples could be shown.

Micro-adaptations are simply particular kinds of animals or plants bringing forth after their own kind. A short-haired dog begetting a long-haired dog would be an example. Dogs producing dogs with genetic variations, which were the result of the sorting or loss of their inherited genetic traits, are simply kinds bringing forth after their own kind, and are both a Biblical and a scientific fact. Biblical Creation meets this key criterion.

Prediction Number Two:

Biblical Creationism predicts that geology will find evidence of a global, catastrophic, aqueous event.

And guess what observable scientific facts reveal?

Science finds that the outer crust of the earth is primarily made of sedimentary layers of rock which were laid down by water. These layers are full of fossils which had to have been buried quickly, before they could rot away or get eaten by scavengers, plus there are countless polystrata fossils which traverse multiple strata layers and attest to the rapid accumulation of the earth's rock layers.

Carbon Dating also lends great support for a recent global flood. Carbon-14, which should decay away in measurable amounts in less than 100,000 years, is found in all fossil-bearing strata layers (evolutionists claim these layers are 580 million years old – more than 5,800 times as long as measurable C-14 should exist). And the C-14 found throughout the various layers is in the same range of amounts.

These findings prove that the earth's strata layers formed recently and during the same event. Only a global flood viably explains this evidence. Biblical Creation meets this key criterion.

So what about the refutability of Biblical Creation?

Though the Bible is not a science book, Scripture claims that plants and animals will bring forth after their kind and that there was a global flood. If the observable evidence does not support these claims, Biblical Creation would be scientifically refuted.

Biblical Creationism is a religious-based theory which is scientifically sound as it meets its own key predictions and it is possible to be scientifically refuted. Furthermore, although it is possible to refute Biblical Creation, it has never been refuted by true scientific research. This is strong evidence that we were created just as the Word of God reports.

In summary:

Darwinism is a religious-based theory that fails its own predictions and is non-refutable.

Intelligent Design is a scientific theory that meets its predictions and is refutable.

Biblical Creationism is a religious-based theory that meets its predictions and is refutable.

Conclusion:

Darwinism does not stand up under the microscope of real science.

Intelligent Design stands up to honest scientific examination.

Biblical Creationism, where testable, stands up to honest scientific examination.

Richard Lewontin, former Harvard Professor of Zoology and Biology stated:

"We take the side of (evolutionary) science ...because we have a prior commitment to materialism. It is not that the methods...of science somehow compel us to accept a material explanation...on the contrary...for we cannot allow a Divine Foot in the door."

From *Billions and billions of demons*; *The New York Review*; January 9, 1997; page 31.

TWENTY

Changing Times

My, how times have changed. Eighty years ago people were called *closed-minded* because they wanted to keep Darwinism out of the public schools. Now the *closed-minded* folks are the Secular Humanists who are dead-set on keeping everything except *millions of years leading to Darwinism* out of those schools.

Real scientific theories come from hypotheses that are built upon observable evidences. Charles Darwin, without any evidence of macro-evolution having taken place, postulated that Biblically correct micro-adaptations would be found to lead to macro-changes. Darwin based his hypothesis on micro-adaptations within the finches he saw on the Galapagos Islands. Of course at that time humans knew nothing about the complexity of the cell, the DNA Code Barrier, Gene Depletion or about the massive genetic information required of such changes.

Since the 1860's everyone has been told that "science proved" Naturalistic Darwinism and that *Matter is all there is* is a fact. Since 1963 – even many years earlier in some countries – this has been taught as if it were the truth in schools around the world.

Secular Humanists have taken over the public educational and scientific establishments, redefining the word *science* to be things pertaining to naturalistic means. They then rule out anything that is non-naturalistic as being *non-scientific*.

So the battle today is not about the evidence, which by far best fits the interpretations of the Biblical worldview. The battle is now over the redefinition of the word *science* which should be "knowledge derived from the study of the evidence."

The last two decades

It's been mainly over the past 20 years that the truth has been seeping out. And today the burden for proof is beginning to fall on the Humanists and this has them beginning to scream. But keep in mind that name-calling and suppression of facts are the last bastions for those who have no evidence to back up their position.

The few brave and honest laborers within the educational and scientific establishments who stand up to the reigning orthodoxy of Darwinism face serious repercussions. Molecular biologist Caroline Crocker of George Mason University is one example. She was barred from teaching certain subjects and consequently released after "mentioning *intelligent design* while teaching her second-year biology course." *Nature*, April 2005.

As intended by the reigning dogma, others skeptical of Darwin's Theory remain fearfully silent.

Even Darwin stated in *Origin of the Species*:

"A fair result can be obtained only by fully stating and balancing the facts and arguments on both sides of each question."

Well, it is long past due that the Naturalistic religious dogma of *millions of years leading to Darwinian evolution* be compared with competing hypotheses and with its own failings.

Humanistic textbooks teach unsuspecting children that there is no reason for their existence and that there is no hope for their future. This is simply another Humanistic lie.

The Creation-Evolution issue is not going to fade away because freethinking people do not accept being told that there are some questions they are not allowed to ask and some answers they are not allowed to question.

As the evidence against Darwinism continues to mount, expect more false claims and hypothetical theories to cover its collapse.

Can anything be done to stem the tide?

As I reported on the first page inside this report, the damage already done to societies around the world by Darwinian philosophy is almost too deep to comprehend and it is almost too late to turn the tide. However, there is a vast difference between *too late* and *almost too late*. So our challenge to you is to pick up your cross and make a difference in this world.

The war is not against flesh and blood, but against powerful forces of darkness.

We hope that this book will lead you to seek a way to make a difference.

Please visit www.Creationministries.org and prayerfully consider involvement with our **Make This Your Own DVD Ministry.** This is the easiest, least expensive and most powerful ministry God has provided for those willing to join us as foot soldiers in this war.

Enlist now, and help us turn the tide.

How does Make This Your Own DVD Ministry work?

One of my God-given gifts is the ability to make fruitful ministry easy for anyone who wants to serve God.

Such a ministry MUST allow you to impact many people for the glory of God and MUST NOT require you to:

A) Speak in front of crowds or
B) Leave your job or
C) Spend large sums of money supporting your ministry.

How can such a ministry be possible?
It's simple.

I believe that God called me to share with others that the Bible is true, word for word and cover to cover. Since I can only reach a finite number of people myself, I asked God to show me how to reach more people for His glory and the answer I received was to begin encouraging everyone to copy and give away CESM DVDs to others.

It really is that simple.

You can help me get the Truth of God's glory to far more people.

This is why I welcome you to Make This Your Own DVD Ministry to copy and hand out life-changing CESM DVDs to as many people as you desire, encouraging them to make copies to give away as well.

Opening the conversation is easy when you combine any of the subjects in the DVDs with, "Would you watch a great DVD if I gave it to you?"

I recommend that you get a set of the CESM DVDs which cover America's Christian heritage, Darwinian frauds, the age of the earth, the global flood, dinosaurs, the formation of Grand Canyon and so much more. Then you can decide which DVD may best influence others. If you prefer to get just one CESM teaching, I suggest you start with either **"50 Facts vs. Darwinism in the Textbooks 101"** or **"If The Foundations Be Destroyed."**

Then obtain your copies to give away by:

a) Making copies on your computer or duplicator;
b) Paying an office supply store to run copies for you;

c) Purchasing BULK RATE DVD sets or singles from us.

Purchasing your CESM DVDs from us will lend CESM some support and allows us to control the quality. But feel free to have your copies made as you choose.

God is raising up many Believers to make it their ministry to hand out copies of CESM DVDs. These weapons of righteousness will tear down the enemy's strongholds while building the foundations for the Gospel of Jesus Christ. This labor of love will yield a tremendous harvest as people copy and spread this knowledge at a grassroots level from person to person, around our cities, country and world.

I invite you to join the hundreds of people who are distributing our DVDs and impacting others with the Truth of God's Word. Let the weak say I am strong! Begin giving away copies of the life-changing CESM DVDs. Your efforts will be part of the great harvest.

You can do this!

I suggest you pray for God's blessings and start looking for a DVD duplicator. Visit our website at www.Creationministries.org and click on DVD ministries for encouragement and helpful ideas on doing this.

1Timothy 6:12

Fight the good fight of faith, lay hold on eternal life, whereunto thou art also called, and hast professed a good profession before many witnesses.

May God bless you and your efforts to serve Him.

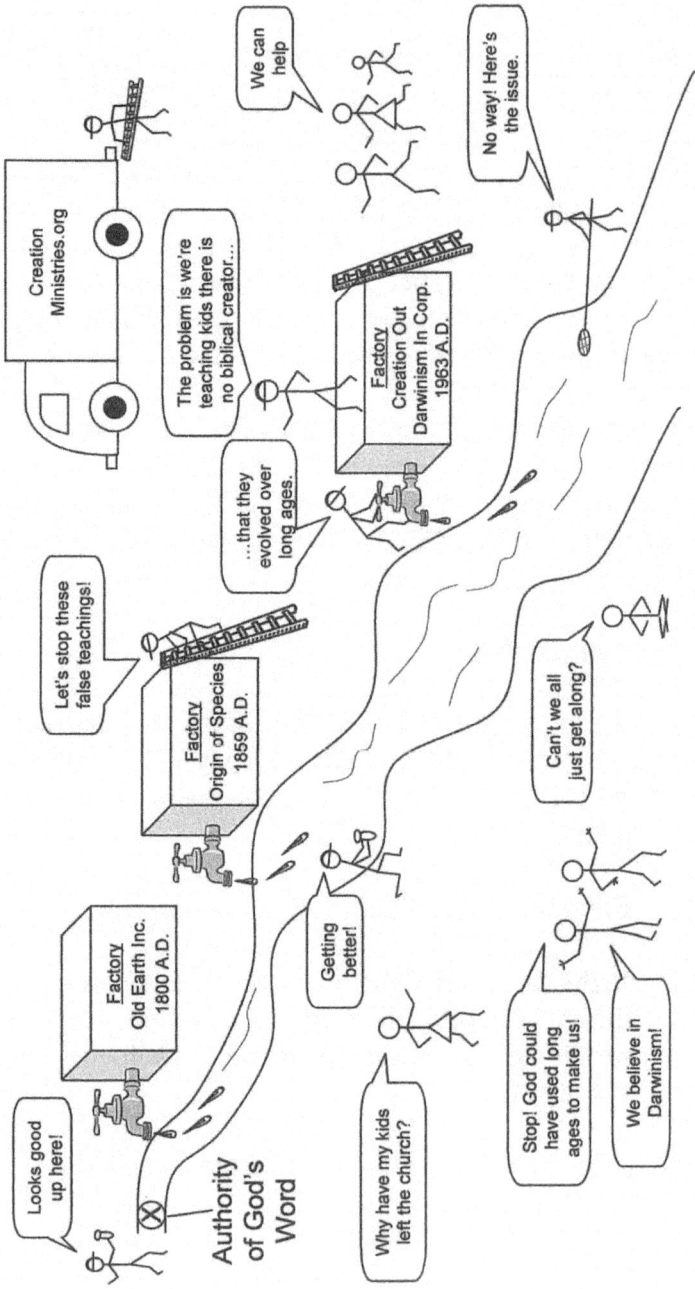

Darwinism leads to compromise – and unbelief in the Bible.

Postscript: What's Ardi got to do with it?

Multiple times during 2009 Darwinian evangelists, backed by their secular allies in the media, announced yet another *undeniable proof* for the foundation of Secular Humanism, which is *millions of years leading to Darwinism.* All previous *undeniable proofs* have been debunked, including *Ida* the lemur and *Lucy.*

Ardipithecus, more simply known as *Ardi,* was hailed as a triumphant proof of evolution.

So what was *Ardi*? Well, decide for yourself based on a few more facts. Although its bones were first found in Ethiopia in 1992, only in 2009 was it being touted as proof for Darwinism. The delay was due to the poor condition of the remains. In fact, a quote from *National Geographic News* reports that the bones were in such bad shape they would turn to dust at a touch.

The folks holding what exists of it say *Ardi is a partial skeleton reconstructed from a **mixture of crushed bones from almost 40 different individuals**.*

This is what the Darwinian-biased scientists *reconstructed* to come up with *Ardi.* In this book you've seen other examples of previous misleading *reconstructions.*

Despite the headline hype, honest scientists have been very skeptical. Anatomist William Jungers of Stony Brook told *NGN*: *"...based on what they present, the evidence for bipedality (walking upright) is limited at best. Divergent big toes are associated with grasping, and this has one of the most divergent big toes you can imagine."*

I am sure that real science will show that Ardi was just an odd ape and no more of a missing link than the many other refuted Darwinian claims.

So what does Ardi have to do with it? Well, it reveals yet again that people must be careful to distinguish real science from biased Darwinian conjecture, because claiming that *Ardi* supports goo-to-the-zoo-to-you evolutionism is nothing more than Humanistic propaganda.

Bookmark these web sites:

The home web site for Creation, Evolution & Science Ministries

www.Creationministries.org

and our partner web site

www.new-earth-thought.com

The GENESIS Heritage Report Series

Go to www.new-earth-thought.com for ordering information and our special package offer, and to sign up to receive FREE Russ Miller's 50 Facts vs. Darwinism; you'll find one fascinating fact in your e-mail inbox daily for 50 days.

Here are the titles of the trade paperback editions in this series:

Report No. 1

The Theft of America's Heritage

**Biblical foundations under siege:
a nation's freedoms vanishing**

Report No. 2

Darwinian Delusion

Exposing the lies that keep on deceiving

Report No. 3

371 Days That Scarred Our Planet

What the stones and bones reveal might surprise you.

Report No. 4

The Submerging Church

Eroded and made irrelevant by compromise

Secular Humanists have brainwashed several generations of unsuspecting children into believing the lies of Darwinism. These children trust what their teachers and textbooks tell them.

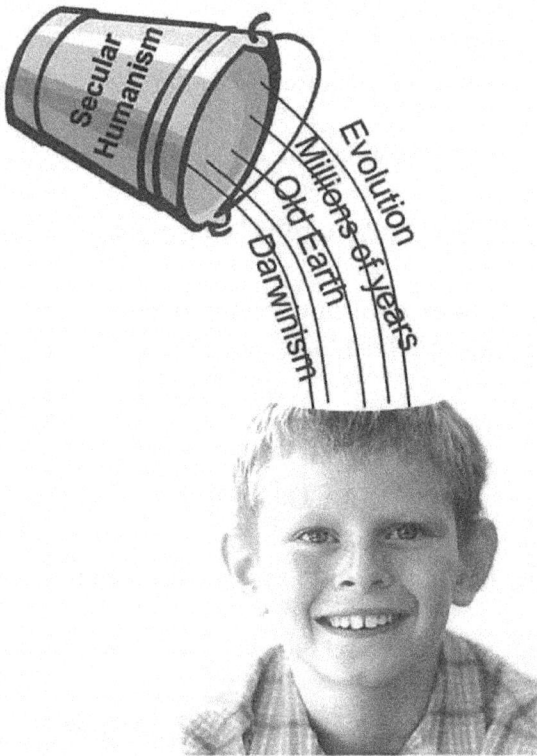

Have you ever wondered if there is anything you can do to help stem the flow of these untruths? If so, please consider Make This Your Own DVD Ministry. Go to www.Creationministries.org and prayerfully consider involvement. Together we CAN make a difference.